Cleaner-Energy Investments

W0036535

Srinivasan Sunderasan

Cleaner-Energy Investments

Cases and Teaching Notes

 Springer

Srinivasan Sunderasan
Verdurous Solutions Private Limited
Mysore
Karnataka
India

ISBN 978-81-322-2953-7   ISBN 978-81-322-2062-6   (eBook)
DOI 10.1007/978-81-322-2062-6

Springer New Delhi Heidelberg New York Dordrecht London

© Springer India 2015
Softcover reprint of the hardcover 1st edition 2015
This work is subject to copyright. All rights are reserved by the Publisher, whether the whole or part of the material is concerned, specifically the rights of translation, reprinting, reuse of illustrations, recitation, broadcasting, reproduction on microfilms or in any other physical way, and transmission or information storage and retrieval, electronic adaptation, computer software, or by similar or dissimilar methodology now known or hereafter developed. Exempted from this legal reservation are brief excerpts in connection with reviews or scholarly analysis or material supplied specifically for the purpose of being entered and executed on a computer system, for exclusive use by the purchaser of the work. Duplication of this publication or parts thereof is permitted only under the provisions of the Copyright Law of the Publisher's location, in its current version, and permission for use must always be obtained from Springer. Permissions for use may be obtained through RightsLink at the Copyright Clearance Center. Violations are liable to prosecution under the respective Copyright Law.
The use of general descriptive names, registered names, trademarks, service marks, etc. in this publication does not imply, even in the absence of a specific statement, that such names are exempt from the relevant protective laws and regulations and therefore free for general use.
While the advice and information in this book are believed to be true and accurate at the date of publication, neither the authors nor the editors nor the publisher can accept any legal responsibility for any errors or omissions that may be made. The publisher makes no warranty, express or implied, with respect to the material contained herein.

Printed on acid-free paper

Springer is part of Springer Science+Business Media (www.springer.com)

Preface

Clean energy is the flavor of the times. With a view to bringing the phrase closer to reality, I would prefer to refer to it in relative terms as *Cleaner Energy*. Investors have had divergent views of the sector: from being a genuine business opportunity, especially when attractive preferential tariffs are on offer, to a peripheral activity that helped "green wash" other business operations. Project financing into Cleaner Energy (often interchangeably used with renewable energy, sustainable energy, or clean energy) has not been intensely researched owing to the specifics of each project, the resource, the geography, the policy environment, the risks involved, and above all, the very availability of relevant information in the public domain. This mandates the study of cases that straddle technologies, financial structures, market conditions, risk factors, and the like. The cases developed herein highlight the relationships among capital structure, managerial incentives, social structures and, most significantly, environmental factors, to help with *Cleaner Energy* project-related decision making, with relevant environmental legislation forming a significant factor in the process.

This compilation of case studies from across countries and cultures involves contemporary technologies including electric vehicles, solar thermal power plants, hybrids, coconut-charcoaling, etc., alongside more traditional and established technologies, viz., wind energy generation and hydro electric power plants. The collection highlights the real-world situations facing individual projects and brings out the strengths and weaknesses of the underlying business propositions. It also throws light on some of the factors that are routinely ignored during project formulation and risk assessment, viz., coordination among public and private agencies, confirmed availability of relatively minor but essential components, possibility of concurrent demand for inputs from different project proponents, etc. Above all, many of the cases studied relate to first-of-their-kind initiatives, successful or otherwise, which command our appreciation all the same for the conviction with which they were taken up for implementation.

This compilation provides a systematic 'guided-tour' of the RE projects for project analysts involving, *inter alia* the development of financial models and culminates with an evaluation of risk and the design of risk-mitigation measures.

It is designed to simultaneously appeal to business school students and to serve as a do-it-yourself guide for practicing executives, policy makers, and consultants.

- The volume is a collection of real-life cases with genuine data sourced from publicly available sources believed to be reliable. Assumptions are made to fill gaps in data, if any, to help with the learning objective.
- The cases are accompanied by teaching notes which carry financial models and other quantitative and qualitative analyses relating to the cases, to help develop multiple perspectives for a given situation.
- The cases cover several countries, currencies, policy environments, technologies, resources and for the benefit of policy makers, consultants, and project analysts and proponents, for them to view *Cleaner Energy* initiatives in a new light.

The case-study method is intended to put the analyst in a specific decision-making environment. When the documented case does not contain all the information needed, it sets the stage for the analyst to seek additional information and perhaps more importantly, to make assumptions when key pieces of the puzzle are not available or readily accessible. The assumptions should be reasonable and consistent with the situation because the "correctness" of the solution derived always depends on the soundness of the assumptions made. Subject to the definition of "reasonableness", a case situation could have more than one "right" solution. It is therefore advisable that the course instructor may be more interested in the analyses and process employed to arrive at the decision than in the arithmetic precision and "correctness" of a solution proffered. The course instructor should also insist on crisp written reports containing key points, the methodology adopted, the analyses, and the conclusions.

A few themes clearly stand out from a detailed analysis of the cases presented herein, a macro analysis of the contexts and a between-the-lines reading of the case documents. The technology employed shall be simple, tested, and proven. Investors routinely face up to market risks and prefer not to absorb technology risks. In other words, mainstream investment projects could employ solar PV and thermal systems, wind energy generators, hydro-kinetic turbines, etc., while relatively nascent technologies such as deep-offshore wind-energy or tidal technologies or other such ventures whose performance and reliability are yet to be proven beyond doubt, have traditionally been funded by governments. The technological spillover thereby is available to society as a whole. In other instances, while the participants chosen could be the best there are, the roles allocated to them might lead to confusion and chaos, as with the Delhi Airport Metro Railway link discussed in this volume.

To the extent possible existing infrastructure—access roads, distribution networks, etc., should be exploited so as to minimize project costs but more importantly, to cut down on implementation time. The San Cristobal wind energy project in the Galapagos Islands illustrates this point admirably well, wherein the implementing agencies were required to build docking piers, roads and other support infrastructure, driving up costs and implementation time. Additionally, it is possible to achieve superior environmental outcomes simply by switching fuels, and conversion technologies, or both, in that order. For instance, existing diesel and gasoline cars could be displaced by electric vehicles as in the Tesla Motors Case or moved on to efficient public transport as demonstrated by the Delhi Airport Metro case.

Involving the communities during the planning stages of the project and taking them into confidence and ensuring that the locals assume moral ownership goes a long way in sustainable operations. *Cleaner Energy* supply should be seen more as a service provision rather than as a hardware lead initiative, requiring constant interaction with the consumers. Presence on the ground and, more significantly, post-installation service support are vital for the sustained operation of plants, across technologies. In the case of the Sri Lanka–REcogen Project, weak social interface had delayed expansion while public opposition to the Norfolk incinerator had almost driven the County Council to bankruptcy. Engaging the local power centers and community leaders, volunteers and NGO workers' helps with coordination and timely implementation, prevents resentment to higher tariffs, increases coordination, and ultimately efficient use of the available energy.

For *Cleaner Energy* technology solutions to be scalable and replicable, capital from private as well as public sources would need to be optimally deployed, structured to suit the specific context, supported by streamlined supply chains for equipment, spares and services, and the institutional framework to manage O&M service delivery and to mitigate the potential risks of O&M deficits. While the promise of power supply raises expectations among the target populations in developing countries, sustained and reliable supply and cost-effective O&M are crucial to supporting local economic development. Grant funding, especially for one-off demonstration projects, has increasingly become scarce as individual projects have not been economically viable or effective in bringing about significant change and measurable development-related outcomes, or both. Replication and the provision of adequate supply of reliable power are hampered by site-specific conditions, including but not limited to, the spatial dispersion of the potential consumers. In all, success with *Cleaner Energy* projects requires a solution-oriented perspective as opposed to a technology and hardware supply motivation.

Transparent and accountable commercialization of energy services is a frequently stated goal of most energy sector reform programs but tariff rationalization in its true form, making the power attractive for private sector investors is without exception impeded by political compulsions. It is believed that the increase in electricity tariffs (or fuels and other intermediate goods) feeds into the general inflation in the economy which is considered a vote-loser for the political decision-makers. Under these circumstances, average tariff realizations are consistently below average costs of delivery. When the transmission and distribution losses and pilferage are added to the mix, the picture becomes less attractive for mainstream private sector investors. In countries with heavy subsidy burdens imposed on the exchequer, the price-cost margin has to be borne by the tax payers, which implies diverting scarce resources from more immediate priorities, say, within the health and education sectors. While rendering decentralized generation less attractive, suppressed retail tariffs also do not encourage investments into conservation or more efficient use of the available electricity.

Supply as well as demand side management should constitute equally important components of energy policy. The use of energy-efficient appliances and the phasing of their use during the day could help optimize sizing of the generation capacities more effectively. Improved matching between generation and consumption could minimize the need for energy storage, round-trip losses, and maintenance requirements. While evaluating the project's ability to service its debt, cash allocations for operation, maintenance, and management should also be integrated into the project's business plan right from inception. Subsidies could be linked to successful and continued operation of the plant and supply of electricity as projected in the business plan.

Generating electricity as a by-product as with the Sri Lanka–REcogen Project, or the proposed Norfolk—Willows incinerator make it more attractive and less risky for investors. The primary product, coconut-char or recycled metal or aggregates that could be used in construction, could be held in inventory and transported across relatively longer distances to earn anticipated returns on investments, while electricity needs to be supplied simultaneous with generation, within the immediate neighborhood of the plant. The San Cristobal wind energy project in Ecuador also serves as a means of increasing the attractiveness of the location as a tourist destination. Such a razor-and-blade (complementary product) business model could help defray tariff risks and potential delays in payments from the utilities.

Customer expectations need to be managed judiciously, by laying out the scope and limitations of the quality and quantity of the proposed service beforehand, to avoid consumer disenchantment and consequent non-payment for the power actually supplied. If for instance, it were proposed that power would be supplied for a few hours each day, for commercial and residential consumers, such timings need to be communicated unambiguously and upfront.

Donor agencies and government departments, when involved, need to assess the project implementation capabilities of the contractors involved in execution to ensure timely completion of projects, avoiding wastage and duplication of effort, and ensuring judicious and targeted use of the funding offered.

Cleaner Energy services function and thrive in an environment with supporting legal and regulatory frameworks and low-cost and effective contract enforcement mechanisms. Simultaneously, public funding continues to play a crucial role in extending utility grids to serve marginal consumers with low incomes, limited consumption, and those who are projected to contribute slim returns to the private investors. For long-term success of electrification programs, therefore, involving mini-grids or extending grids to underserved clusters, both the business case and the development argument need to be strong and compelling. Public and donor funding could be used to close the "viability gap" and to access the "last mile".

Irrespective of the price paid for substitutes, viz., candles, dry cells, kerosene lamps, etc., when customers are required to pay for electricity, they are bound to compare the service and costs available to grid connected consumers—the source of generation whether diesel or solar PV is a matter of irrelevant detail as far as consumers are concerned, and rightly so—and would expect *not* to pay a premium over grid-connected customers next door. Frequently, inexperienced third-party agencies plan and implement projects with the implicit assumption that consumers would be willing to pay higher tariffs for superior illumination, relative to the candles and lamps displaced: independent of tariffs paid by grid-connected acquaintances residing in the vicinity.

Once an irreversible investment is made and a cleaner energy project asset is installed, it needs to provide the intended service and earn projected revenues, and shall not be stranded by subsequent developments as with the expansion of the utility grid supplying, possibly unreliable, but potentially cheaper power; adequate safeguards and guarantees need to be built into concession agreements to protect investor interests.

Lifecycle costs of electricity generated (levelised cost of energy generated—LCoE) are frequently employed as benchmark figures for policy formulation. This holds good only if the terms of financing for given projects match revenue generation profiles, for mutually exclusive projects to be rendered comparable. While fossil fuel-based generation plants are exposed to fuel price risks, renewable energy plants, viz., solar PV plants and wind energy generators face uncertainties relating to the sustenance of preferential tariff regimes. The concept of energy security therefore needs to be broadened to include impacts of fossil fuel price variations on the economy while comparing the two alternative courses of action. The challenge therefore lies in minimizing project-level uncertainties while ensuring predictable and reliable energy supplies. Long-term policy stability is fundamental to delivering such levels of deployment of energy efficient end-use and renewable energy technologies and to ensuring that these technologies continue to operate and deliver services as planned.

Efforts to 'decarbonise' the economy and to provide reliable electricity for productive use are a worthwhile investment, simultaneously curtailing runaway climate change. In the context of several developing countries, rural electrification, cleaner energy use, poverty alleviation, ecosystem protection, watershed management, combating desertification, and achieving social equity are all inter-related goals.

The case documents and teaching-notes are intended to help instructors guide classroom discussion on the underlying issues and are definitely not meant to serve as an endorsement, or to judge regulatory prescriptions, management styles, decision-making, or outcomes. The cases and teaching notes are designed to provide multiple perspectives on a given situation, to aid academic discourse, and to help evolve alternative recommendations subject to relevant assumptions relating to the future.

Mysore, India, June 2014   Srinivasan Sunderasan

Contents

Abbreviations

CDM	Clean Development Mechanism
CSP	Concentrating Solar Power
DNI	Direct Normal Irradiation
EfW	Energy from Waste
EIB	European Investment Bank
EPC	Engineering, Procurement and Construction
EWEA	European Wind Energy Agency
FERC	Federal Energy Regulatory Committee (USA)
GHG	Greenhouse Gas
GWEC	Global Wind Energy Council
HTF	Heat Transfer Fluid
IFC	International Finance Corporation (member World Bank Group)
IRR	Internal Rate of Return
kV	kilo Volt
kW	kilo Watt
kWh	kilo Watt hour
LCoE	Levelized Cost of Energy
MW	Mega Watt
MWh	Mega Watt hour
NGO	Non-governmental Organization
NPV	Net Present Value
O&M	Operations and Maintenance
PLF	Plant Load Factor
PPA	Power Purchase Agreement
PV-T	PV–Thermal (hybrids)
RD&D	Research, Design and Development
SHP	Small-hydro Power
SPV	Special Purpose Vehicle

t CO_2e	ton of Carbon-di-oxide equivalent
UNDP	United Nations Development Program
UNFCCC	United Nations Framework Convention on Climate Change
WAPP	West African Power Pool

About the Author

Srinivasan Sunderasan is an Economist at Verdurous Solutions Private Limited (Mysore, India), which is an investment advisory and consultancy specializing in management and financing aspects of renewable energy, microfinance, water, waste, and other 'sustainable development' initiatives. Prior to this, he was the Deputy Country Manager (India) of Photovoltaic Market Transformation Initiative (PVMTI), an Investment Officer with the South Asian Region for Solar Development Group (SDG) and Triodos Renewable Energy for Development Fund. He has obtained his Ph.D. from University of Vienna, Austria in 2005, specializing in business economics.

He has 20 years of rich and diversified experience as a techno-commercial professional, in planning, project management, research, academics, and consultancy in various industries including construction, telecom, energy, financial services, and 'sustainable development'. He is recognized as an expert on rural energy supply models by the International Solar Energy Society and a specialist in evaluation and financing for renewable energy and other cleantech ventures. He has authored, and contributed to, many books on related subjects published by Routledge, Taylor & Francis Group, Nova Publishers, and Springer Verlag. He is an editorial board member of *J. Reviews on Global Economics* and expert peer-reviewer with prestigious publishers including Elsevier B.V. and Springer Verlag. He is guest faculty of finance/economics at reputed business schools.

Chapter 1
Solar PV–Thermal Hybrids: Energy in Synergy

Valuing a Hybrid Technology Package

> *PV–T is a holistic energy solution that suits the needs of the market, so it really has legs.*
> —Daniel Barber, Director, Solimpeks Australia

> *We're not here to take over the PV or solar thermal market. We're creating a new market space. We are the third option.*
> —Toby Greenane, Co-Director, Solimpeks Australasia.

Background

What could possibly lead to a whole new way of viewing rooftop active solar technology deployment went unnoticed. Almost. Turkish solar thermal system manufacturer *Solimpeks* (www.solimpeks.com) had been manufacturing solar thermal collectors, insulated water tanks, and system accessories for close to four decades. Over the years, the company had expanded research and production into high-quality solar thermal collectors, thermo-siphon systems, and mounting assemblies. The company had opened its second manufacturing plant in Kenya as recently as 2011 and enjoyed market presence in over 70 countries worldwide. The Turkey/Izmir head quartered company also built solar thermal system components and assemblies on behalf of several European vendors, with its product retailed under several reputed brands.

More significantly, the company had been developing solar PV–thermal (PV–T) hybrid systems for almost a decade. PV–T hybrid modules brought traditional solar photovoltaic panels and solar thermal collectors together into a single package. The combination facilitated the capability to generate electricity from solar insolation along with the ability to heat water for domestic or commercial use. The hybrid system generated both electricity and usable heat from the same panel in different proportions, subject to design specifications as shown in Fig. 1.1.

© Springer India 2015
S. Sunderasan, *Cleaner-Energy Investments*,
DOI 10.1007/978-81-322-2062-6_1

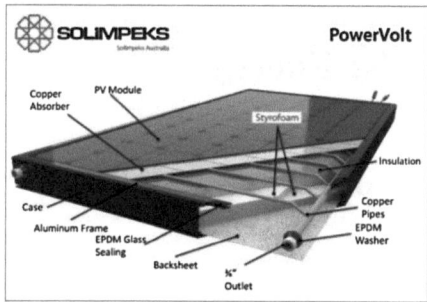
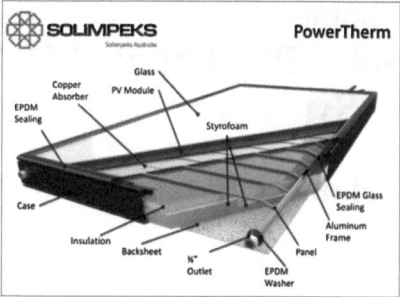

Fig. 1.1 Solimpeks *powerVolt* with electricity as the primary output and *powerTherm* modules when heating water was the primary goal (*Figure credit* Solimpeks marketing brochures/product data sheet; reproduced with permission)

Technical Overview

Solar PV cells worked on solar radiation but degraded at higher temperatures. For each degree rise in ambient temperature above 25 °C, the traditional solar PV panel's efficiency was known to be lowered by 0.4 % (Kyriakou 2014). In the PV–T panel, the efficiency of power generation from the solar PV component was optimized better, as the surplus heat was transferred into heating water. The withdrawal of heat from the PV panel through the cooling and insulating effect of water meant that the semiconductor cells operated within a narrower temperature range and hence suffered lower degradation over time. This helped generate 15 % more power than standard solar PV modules of similar power rating for a given location. Simultaneously, the withdrawn heat was put to good use through the thermal collectors. Support structures were simpler and installation was relatively straightforward, and the combined system required lesser roof space. The energy and heat output ideally suited domestic and commercial applications, clinics, hospitals, swimming pools, restaurants, schools and hostels, food processing and packaging units, etc.

The "PV-biased" *PowerVolt* modules were designed for electricity as the primary output, while the "heat-biased" *PowerTherm* modules were focussed on heating water with electricity as a secondary yield (Solimpeks 2014). PowerVolt modules were rated at 200 Wp for the PV component and could heat 65 l of water to 60 °C in one hour (~619 W of thermal output).[1] The PowerTherm modules were rated at 180 Wp (PV) and could heat 65 l of water each hour to a maximum temperature of 70 °C. Water flow rates through the modules could be adjusted to ensure the right amount of heating, commensurate with prevailing weather

---

[1] "Solimpeks Makes Australian History Again With Its PV–Thermal Hybrid Solar Panel", solimpeks.com.au, posted 19 September 2013, http://solimpeks.com.au/solimpeks-makes-australian-history-again-with-its-hybrid-pv-thermal-solar-panel/, last accessed 3 April 2014.

conditions. The hybrid modules were mounted and installed exactly along the lines of traditional systems, and hence, experienced installers could work with the systems with existing skills and available tools.

The 14 m^2/2 kWp PV–T installation in Australia/New South Wales (NSW)/ Killara was intended more as an experimental setup to test the hybrid system and to collect performance data in real-world conditions. The ten panels were arranged in five columns and two rows, with each module in "landscape" mode to maximize heat evacuation through the string. The installation included a pyranometer (to collect solar radiation data), anemometer (to measure wind speeds), and equipment to measure ambient temperatures, in/out water temperature, and PV energy generated (Morris 2013). *Solimpeks* dealers were often taken to the site to demonstrate the hybrid system's functioning and more importantly to convince them of the simplicity of the design and installation.

The Killara installation was the first time ever that *Solimpeks* had used the 200 Wp hybrid panels and was the first accredited hybrid PV–thermal solar system in Australia. It was reported that the customer subsequently requested the company to double the installed capacity to 4 kWp (Solimpeks Information Sheet 2014).

Project Cost and Means of Finance Project

The *PowerVolt* module was estimated to generate 328 kWh/m^2/year of heat and 216 kWh/m^2/year of electricity (Solem Consulting 2014). At 14 m^2, the total annual output from the installation would be 7,616 kWh, representing an efficiency of 43.47 %—higher than any stand-alone solar PV system but lower than a solar thermal water heating system. Subject to the tariff plan selected,[2] electricity tariffs could range between AUD 0.28 and 0.32 per kWh and monthly power expenses between AUD 127 and AUD 142 for "medium usage" for a household without solar PV panels. At an average power tariff of AUD 0.30 per kWh, 7,616 kWh would represent an annual saving of AUD 2,285. But the question yet left unanswered was whether the hybrid system was cost-effective relative to (i) stand-alone solar PV and solar thermal systems and (ii) grid supplied electricity.

---

[2] https://www.energyaustralia.com.au/quote?postcode=2000&customerType=RES#!chooseplan, last accessed 5 April 2014.

Select Input Data to Estimate Project Returns

Parameter	Unit	Value	Unit/Remarks
Solar PV part of hybrid system	kWh	216	Per m^2
Solar PV part of hybrid system	kWh	3,024	Per 14 m^2
Stand-alone solar PV system	kWh	3,478	Per 14 m^2; 15 % additional sizing required
Stand-alone solar PV system	kW	1.932**	At 6.0 kWh/kWp for 300 days/year
Cost of stand-alone PV system	*AUD*	*9,660***	*At AUD 5 per installed Wp*
Solar thermal part of system	kWh	328	Per m^2
Solar thermal part of system	kWh	4,592	Per 14 m^2
Solar thermal part of system	kWh	5,281	Per 14 m^2; 15 % additional sizing required
Solar thermal part of system	kW	2.94**	At 6.0 kWh/kWp for 300 days/year
Cost of stand-alone PV system	*AUD*	*14,669***	*At AUD 5 per installed Wp*
Total system cost: (stand-alone solar PV + stand-alone solar thermal)	AUD	24,329**	
Annual saving on electricity consumption	AUD	2,285	Not considering incentives available for solar PV installations

**Author's estimates/computations

Teaching Note

Case Synopsis

The case deals with the integration of solar PV and thermal systems. Efficiency of solar PV systems degraded as ambient temperatures rose. The withdrawal of generated heat helped cool and insulate the solar PV component and hence to improve electricity generation by over 15 % on average. The solar water heating system that evacuated the heat was a string of collectors placed on the back of the PV layer and connected in "series." The hybrid collectors were placed in "landscape" mode as opposed to the more traditional "portrait" format to assist with the thermo-siphon within each such hybrid module.

For about 40 years, *Solimpeks* of Turkey had been continuously improving upon solar thermal collectors and had been producing system components and assemblies in ever larger numbers. Its product was sold in over 70 countries under its own label as well as under several internationally reputed brand names. The case employs a 2 kWp solar PV–thermal hybrid system from Australia, New South Wales, Killara to illustrate the economics of the hybrid system.

The company believed that hybrid PV–thermal technology provided the benefits of both solar PV and solar thermal, conserved on installation effort, mounting structure, and rooftop space. The company asserted that it was creating a new market space for its own product, as well as for the competition that would almost inevitably follow, if the product were to succeed. The case therefore calls for a validation of the hybrid system's costing in relation to the sets of stand-alone systems being displaced, as well as in relation to utility supplied electricity/gas.

Case Question

To test for the financial viability of the hybrid solar PV–thermal system with reference to stand-alone solar PV and solar thermal systems and to utility supplied electricity/gas, etc.

Teaching Objectives

- Enabling students to develop an understanding of the scenario involving the introduction of a new product/technology.
- For students to view the product and its deployment from the customers' perspective and to compile a simple financial model with available data in the manner of a rational prospective customer.
- Preparing an elementary NPV/IRR model and undertaking sensitivity and scenario analyses.

- Developing a set of price, interest cost, and performance scenarios that would help the product/technology compete against its nearest rivals.

Case Objectives and Use

The primary objective of the case is to help course participants visualize and define the benchmarks to compare the performance and pricing of a new product/technology. For the present case, data required for the comparative analysis of alternatives are provided. However, the case is intended to provide students with an insight to the data set to be assembled for similar analyses of other technology packages. Analysts are then required to organize an elementary albeit, robust, financial model to compare the returns on each of the alternatives available. Indirectly, this model could be used to arrive at first costs, interest rates, and performance parameters for the new product/technology.

In the present case, the pilot project was implemented purely as a proof-of-concept and to demonstrate the system's functioning. Yet, the company expected to make rapid inroads into the Australian and other markets with the hybrid product. Cost overruns during the implementation of the experimental/demonstration setup might not impact the product's competitiveness or the company's profitability very significantly. But implementing a well-considered pricing strategy is always paramount to medium-term survival and long-run success. The alternative scenario would have been for the customer to install stand-alone solar PV and solar thermal systems to generate electricity and to heat water. In turn, each of these would have been benchmarked against the price of electricity/fuels delivered by existing utilities. Hence, despite the company's stated position of not competing against stand-alone systems, in reality, the rational customer would benchmark the price performance parameters of the hybrid system against them.In turn, the decision to invest in stand-alone solar PV and solar thermal systems would by itself, have been made after taking into account the costs of utility delivered energy services.

The case exposes analysts to a customer's decision-making situation, where alternative technology packages sought to compete to provide undifferentiated services: electricity and hot water. Managers would be required to arrive at a pricing strategy that would balance between the additional investments into the combined package and the incremental *tangible* efficiencies it delivered. Improved aesthetics could serve as a value enhancement and as an additional, possibly non-monetary benefit.

Teaching Plan

The case presents a product introduction strategy wherein the agency concerned was required to balance between incremental aesthetics, efficiency, and cost benefits of the new technology package, against the additional cost to the consumer.

The first step would be to identify the nearest competitors to the technology package poised for commercial launch. Since the new technology package was a solar PV–thermal hybrid, the incumbents are readily identified: stand-alone solar PV packages and solar thermal water heating systems. A small segment of the market might have been willing to pay a premium to lead with the deployment of the new technology, for its lower rooftop area coverage or for the aesthetic benefits it delivered or a combination thereof. For a mass rollout, the company would need to demonstrate that the incremental first costs, if any, were compensated by additional financial benefits—increase in electricity output and superior monitoring of water flow and temperature.

The instructor could guide the course participants to analyze the financial model for two alternative scenarios (i) computing the first cost of the stand-alone substitutes delivering comparable benefits and (ii) arriving at price–performance–returns scenarios that would help with widespread acceptance of the new technology. The instructor could then guide the discussion on enhancing returns on research and development through introductory premium pricing before the competition managed to replicate, and to possibly improve upon, the product.

The instructor could guide the students to evaluate the project in a number of small steps as laid out herein below:

- Table 1—The solar PV equivalent of the new technology package.
- Table 2—Solar PV stand-alone system sizing and costing.
- Table 3—Solar thermal (equivalent) stand-alone system sizing and costing.
- Table 4—Annual savings from displacing electricity/gas/other fuels (for the PV–T hybrid system or for two stand-alone systems with comparable service output).
- Table 5—Benchmark cost of stand-alone solar PV and solar thermal systems and returns on the investment over a 20 year investment horizon.
- Table 6 (and graph)—Solar PV–T hybrid system cost and return scenarios, holding performance parameters constant.

What Happened Next

In July 2013, *SunDrum Solar LLC* (www.sundrumsolar.com) announced that it had achieved a one-hour peak delivery record of 86 % efficiency for a hybrid solar PV–T system. The feat was achieved between 1,400 and 1,500 h when 870 W of thermal energy and 200 W of electrical energy were delivered by each solar PV panel—SunDrum collector combine. Standard solar thermal systems consumed about 60 % of the incident energy to heat water, while traditional solar PV systems had achieved 19 % conversion efficiencies on the field. An output that was greater than the sum of both efficiencies was a record for any fixed, non-tracking hybrid array SunDrum Solar (2013).

The financial projections for the project would need to be drawn-up one step at a time:

1. Solar PV–thermal hybrid system outputs (2 kW Pilot Project in New South Wales, Australia)

Electricity generation per sqm per year	kWh	216
Electricity generation per 14 sqm per year	kWh	3,024
Thermal generation per sqm per year	kWh	328
Thermal generation per 14 sqm per year	kWh	4,592

2. Solar PV stand-alone system sizing and price comparison

Required capacity	15 % over hybrid system	3,478
Electricity generated per day*	kWh/kW installed	6.00
Operating period per year*	Day	300
System size	kW	1.932
System unit cost*	AUD/Wp	5.00
System cost	AUD	9,660

3. Solar thermal stand-alone system sizing and price comparison

Required capacity	15 % over hybrid system	5,281
Electricity generated per day*	kWh/kW installed	6.00
Operating period per year*	Days	300
System size	kW	2.934
System unit cost*	AUD/Wp	5.00
System cost	AUD	14,669

4. Savings in electricity/fuel expenses

Electricity tariff*	AUD/kWh	0.30
Electricity consumption displaced per year	kWh	7,616
Savings achieved for years 1–10	AUD	2285.00
Savings achieved for years 10–20 (@90 %)	AUD	2057.00

* Author's assumptions

5. Total first cost of stand-alone solar PV and solar thermal systems

1.93 kWp solar PV system	AUD	9660.00
2.93 kWp (eq) solar thermal system	AUD	14669.00
4.866 kWp equivalent system**	AUD	24329.00

** Author's computations

6. Returns earned on stand-alone solar PV and solar thermal systems

0	1	2	3	-	10	11	12	-	18	19	20	
(24,329)	2,285	2,285	2,285	-	2,285	2056.5	2056.5	-	2056.5	2056.5	2056.5	6.485 %
First cost of stand-alone systems	Annual savings from electricity (equivalent) displaced in years 1–10					Annual savings from electricity (equivalent) displaced in years 10–20						IRR on investment

7. Returns on solar PV–thermal hybrid systems, holding performance parameters constant

System cost (AUD)	IRR (%)
21647.00	8.05
23308.00	7.05
25109.00	6.08
29815.00	4.00
35704.00	2.00
43385.00	0.00

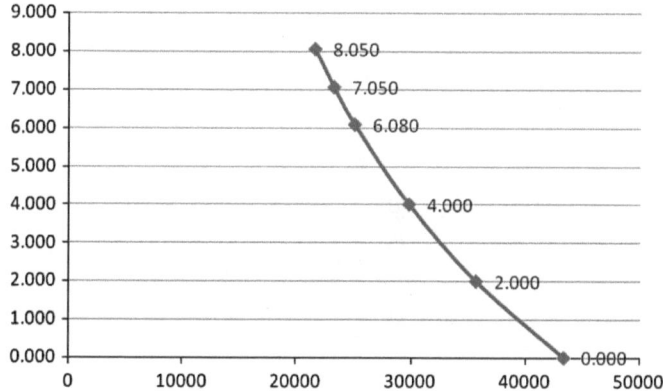

Graph shows PV–T hybrid system costs on the horizontal axis and expected returns on the vertical axis, holding other parameters constant.

References

Kyriakou D (2014) Solimpeks Hybrid PV-T Technology. *Electrical Connection*, 18 March 2014. http://electricalconnection.com.au/article/10021228/solimpeks-hybrid-pv-t-technology. Accessed 3 Apr 2014

Morris N (2013) World First In Australia for PV Thermal Hybrid Technology, *Renew Economy*, 7 Feb 2013. http://reneweconomy.com.au/2013/world-first-in-australia-for-pv-thermal-hybrid-technology-24218. Accessed 3 Apr 2014

Solem Consulting (2014) Theoretical Performance Assessment of Solimpeks PV-T Solar Collectors, 19 Apr 2012, p. 4 of 34. http://solimpeks.com.au/downloads/Solem-Solimpeks%20PVT%20performance%20report.pdf. Accessed 4 Apr 2014

Solimpeks (2014) Solimpeks Hybrid PV-Thermal Solar Panels in Australia (2014) http://www.solarchoice.net.au/blog/solimpeks-hybrid-pv-thermal-solar-panels-in-australia/. Accessed 3 Apr 2014

Solimpeks Information Sheet (2014) Hybrid PV-Thermal Case Study 1: Killara, December 2013. http://infinitypower.com.au/wp-content/uploads/2013/12/Solimpeks-Case-Studies-HiRes-Edit.pdf. Accessed 4 Apr 2014

SunDrum Solar (2013) SunDrum Solar Hybrid PVT System Reaches Peak 86 % Delivery Record, Renewable Energy Focus, 9 July 2013. http://www.renewableenergyfocus.com/view/33360/sundrum-solar-hybrid-pvt-system-reaches-peak-86-delivery-record/. Accessed 5 Apr 2014

Chapter 2
The Condit Dam: A Make-or-Break Decision

Discounting Cash Flows and Comparing Alternatives

You hate to see it go; it's good, carbon-free energy.
—Tom Hickey, senior engineer for hydro resources
for PacifiCorp

We might have different ideas of what's going to happen next.
—Susan Hollingsworth, local rafting instructor

Introduction

The White Salmon River is a glacier-fed river originating on the slopes of Mount Adams and joining the Columbia River downstream of the Condit Dam. At the time that a decision was due, in the year 1996, the Condit Dam, featured in Fig. 2.1, across the White Salmon River in the US state of Washington was the largest dam ever considered for potential dismantling and removal.

The area was known for its scenic natural beauty and as a site for recreational activities such as white-water rafting and fishing.[1] The 125-ft-high dam itself had been a historic engineering and architectural landmark for close to a century, since impoundment of water began in 1911, and had been supplying electric power to local industry and to the city of Portland, Oregon State.

In the year 1947, PacifiCorp (www.pacificorp.com), presently owned by Mid-American Energy Holdings Co., a unit of Warren Buffet's Berkshire Hathaway Inc. (Dininnyj 2012), acquired the project with its 14.7 MW of installed power generation capacity with average annual generation of 79,700 MWh (79,700,000 kWh or "units") of actual generation. Sale of electric power has been yielding annual revenues of USD 2,896,000, while costing USD 40,000 per annum in operating expenditure.

The Condit Dam was designed and built with fish passage systems ("fish ladders", illustrated in Fig. 2.2) which were destroyed by floods shortly after the dam

---

[1] http://www.pacificorp.com/es/hydro/hl/condit.html last accessed 19 October 2012.

© Springer India 2015
S. Sunderasan, *Cleaner-Energy Investments*,
DOI 10.1007/978-81-322-2062-6_2

Fig. 2.1 The Condit Dam: 90 foot wide at the base and 92 acre Northwestern Lake reservoir behind; photo courtesy PacifiCorp: reproduced with permission

Fig. 2.2 Fish ladders enable fish to pass around barriers by swimming and leaping up a series of relatively low steps into the waters on the other/upstream side. *Figure Credit* National Oceanic and Atmospheric Administration, US Department of Commerce

was commissioned for service. Thereafter, to compensate for the obstruction caused to the natural upstream migration of Columbia River Chinook salmon and disturbance caused to steelhead habitat, the project was to develop fish hatcheries. Salmon were also manually trapped and transported upstream.

Environmental Credentials and Concerns

Extraction of power from falling water, "hydropower" has been one of the oldest sources exploited, and among the lowest marginal cost technologies of generation. The environmental impact of hydropower projects had, however, been subject to intense debate over the years. Large hydropower involving realigning of river systems and impounding of water was slated to have serious adverse effects on existing marine ecosystems, as well as on the newly submerged lands. Egypt's Aswan Dam, for instance, was confirmed to have caused a certain degree of environmental problems over its 25-year existence (Rashad and Ismail 2000).

However, the United Nations Framework Convention on Climate Change (www.unfccc.int) recognized Brazil's 3,750-MW Jirau hydropower plant under the Clean Development Mechanism (CDM), making it the world's largest "renewable energy plant" to be registered under the Kyoto flexibility mechanism (Brazil's Jirau Hydro Project World's Largest CDM-Registered Renewable Plant 2013).

Small hydroprojects, on the other hand, were likely to leave small and localized impacts. But, small dams contributed to environmental degradation when such microeffects accumulated across multiple dam sites. It was possible that the hydrologic and habitat change and other biophysical impacts of small hydropower schemes could exceed those of large hydropower projects when such effects were normalized for plant size (per MW) of power generated (Kibler and Tullos 2013).

In 1993, the Federal Energy Regulatory Commission (FERC), in keeping with National Marine Fisheries Service standards, called for the installation of modern fish ladders and other modifications to the Condit Dam project/operations to allow for a 95 % survival rate of migrating salmon (Federal Energy Regulatory Commission 1996).

Preliminary estimates indicated that the modifications would require a capital investment amounting to USD 60 million while also reducing the amount of water available for power production.[2] Analysts could consider ten additional years of the project.[3] The costs of decommissioning the project and consequent removal, clean-up and rehabilitation to be borne by PacifiCorp were estimated at USD 37.0 million.[4]

---

[2] PacifiCorp, *Condit Hydroelectric Project, The Relicensing Process and Decommissioning.*

[3] Author's insertion to define a time horizon for project decision-making.

[4] http://news.nationalgeographic.com/news/2011/10/111028-condit-dam-removal-video/ last accessed 19 October 2012.

Additional Information

- The company expected to be in a position to mobilize capital at an average cost of 3.5 % per annum in nominal terms.[5] Subsequently, the water available for power generation was estimated to provide for 70 % of the average annual generation that was achieved prior to the installation of the fish ladders. The average tariff receivable was slated to remain the same for the first year while increasing at US 0.1 cent each year thereafter for the next 9 years.
- The compound (dam plus fish ladder) project was expected to generate incremental revenues from tourism (camping, fishing, rafting) and support services amounting to US$ 0.25 million in year 1, growing at 2 % per annum thereafter.
- Operating expenses were estimated to rise to USD 45,000 in year 1 and further increase at 2 % per year over the previous year's expenses, for the next 9 years. PacifiCorp did not anticipate expanding its administrative setup, but US$ 0.25 million of administrative expenditure was to be allocated to the modified project.

---

[5] Author's estimates.

Teaching Note

Case Synopsis

The environmental impacts of hydroprojects—large and small—have been the subject of debate for several decades now. Researchers have pronounced different verdicts at different points in time. Some have confirmed that the realignment of river flows, and construction of dams to impound large quantities of water, had led to adverse consequences to natural ecosystems. Others have opined that the cumulative impact of a large number of small projects could have more serious consequences per megawatt (MW) of energy extracted, relative to larger projects.

The case is a singular exposition of the conflict between environmental activism and business interest, with project analysts having to make a make-or-break decision, quite literally. Regulators had provided two choices to PacifiCorp: to build fish ladders to help certain species of fish migrate along the White Salmon River from downstream of the Condit Dam to the Northwestern Lake reservoir behind the dam, or to completely dismantle the dam itself. The case provides analysts with details of various elements of cash flows for either scenario and calls for a reasoned analysis and a recommendation.

Case Question

Would it be more prudent to build the fish ladders and conform to operating standards prescribed by the regulators, or would it be more viable to dismantle the Condit Dam entirely.

Teaching Objectives

- Enabling students to prepare cash flow statements, taking cognizance of the figures provided within the case, and recognizing the "incremental" cash flows alone.
- Providing students with the experience of undertaking a microeconomic analysis of a real-world situation, and to quantify future states of the world for each of the alternatives.
- Enabling students to undertake a sensitivity analysis to come up with a range of prescriptions on the ultimate decision.
- Enabling students to undertake a scenario analysis, *inter alia*, listing the range of values those individual variables could assume, to highlight the relationships among variables being analyzed, and to arrive at the most viable prescription.

Case Objectives and Use

The case deals with the analysis of cash flows and provides students with a binary option: to continue operations or to terminate operations and dismantle the project asset. Among other things, it provides students with exposure to decide on the elements of cash flows that should be included in such an analysis and, more specifically, on the "non-incremental" cash flows that need to be ignored.

The instructor could choose to expand the scope of the case to include secondary aspects that could impact the decision and on whether PacifiCorp could engage in some form of complementary product (razor and blade) pricing strategies. In other words, the instructor could hint at investing in a presently loss-making venture with the prospect that the investments could be recovered by secondary activities, viz., tourism, etc. Alternatively, the instructor could open up a discussion on whether the investment could help PacifiCorp raise electricity tariffs owing to its newly acquired environment-conscious image. At this stage, the differences between consumers' stated and revealed preferences could be highlighted.

Teaching Plan

The case presents a classical microeconomic bargaining between two alternative and mutually exclusive scenarios. Regulators and government agencies had attempted to balance commercial interests with the preservation of the natural environment. The conflict between, or the overlap of, the two spheres could be brought up for discussion.

Scenario 1: Terminating business operations and dismantling the Condit Dam

$$\text{Projected one-time expenditure (outflow)} = \text{USD } 37.00 \text{ million}$$

Scenario 2: Building fish ladders and continuing operations for ten years

$$\text{Initial investment into construction of fish ladders} = \text{USD } 60.000 \text{ million}$$
$$\text{Present-Value of stream of EBIT cash-flows} = \text{USD } 20.672 \text{ million}$$
$$\text{Net Present Value from building fish-ladders (net outflow)} = \text{USD } -39.327 \text{ million}$$

Given the input parameters, it is evident that dismantling the dam is less expensive relative to building the fish ladders and extending the life of the project asset by a period of 10 years.

What Happened Next

PacifiCorp decided to remove the dam rather than construct the fish passage as mandated by the regulatory authorities. The dismantling of the dam commenced past noon on October 26, 2011. The draining of the sediment retained by the reservoir turned the downstream waters turbid. The removal of the dam was completed by mid-2012, and for the first time in a century, migrating steelhead and other species of fish were sighted upstream of the former dam site.

Based on a background search, the case could be enlarged to study:

- The range of values the variables such as power tariffs, interest rates, etc., could take.
- Relationships among sets of variables: power tariffs and quantum of power sold; interest rates and project NPV; numbers of tourists, per capital spent by tourists and total revenues from tourism.
- Combined with a scenario analysis from 2 above, the impact of an extended tenure of the concession would make: if the concession were to have been extended by 5/10 years.
- Other sources of income from the combined dam + fish passage project.
- The ideological dichotomy: should power tariffs be raised to cover the cost of the fish ladders (power consumers pay for the environmental impact) or should the government bear the burden of the capital expenditure in the spirit of the environment being a public good (all tax payers contributing to make good the adverse environmental impact of the dam).

The cash flow chart for the project would need to be drawn up one step at a time.

1. Electric power tariff earned at present

Annual generation of electric power	kWh	79,700,000
Annual revenues	USD	2,896,000
Tariff	USD/kWh	0.03633626

2. Projected quantum of power generation

Annual generation of electric power	kWh	79,700,000
Proportion of generation when fish ladders are built	%	70
Projected generation when fish ladders are built	USD/kWh	55,790,000

3. Projected annual tariff

Year	1	2	3	4	5	6	7	8	9	10
Tariff/ kWh	0.0363	0.0373	0.0383	0.0393	0.0403	0.0413	0.0423	0.0433	0.0443	0.0453

4. Revenue model: projected revenue from the sale of electric power

Year		1	2	3	4	5	6	7	8	9	10
Power generated	kWh	55,790,000	55,790,000	55,790,000	55,790,000	55,790,000	55,790,000	55,790,000	55,790,000	55,790,000	55,790,000
Tariff	USD/ kWh	0.0363	0.0373	0.0383	0.0393	0.0403	0.0413	0.0423	0.0433	0.0443	0.0453
Revenue	USD	2,027,200	2,082,990	2,138,780	2,194,570	2,250,360	2,306,150	2,361,940	2,417,730	2,473,520	2,529,310

5. Other income: Income from tourism escalated at 2 % per annum

Year		1	2	3	4	5	6	7	8	9	10
	Escalation										
	USD	2 %									
Revenue from tourism		250,000	255,000	260,100	265,302	270,608	276,020	281,541	287,171	292,915	298,773

6. Total revenue, operating expenditure, and EBIT

Total revenue	2,277,200	2,337,990	2,398,880	2,459,872	2,520,968	2,582,170	2,643,481	2,704,901	2,766,435	2,828,083
Operating expenses	45,000	45,900	46,818	47,754	48,709	49,684	50,677	51,691	52,725	53,779
EBIT	2,232,200	2,292,090	2,352,062	2,412,118	2,472,259	2,532,487	2,592,803	2,653,211	2,713,710	2,774,304

Important to ignore allocated head office/administrative costs that are not incremental

References

Brazil's Jirau Hydro Project World's Largest CDM-Registered Renewable Plant (2013) http://www.hydroworld.com/content/hydro/en/articles/2013/06/brazil-s-jirau-hydro-project-world-s-largest-cdm-registered-rene.html. Accessed 15 June 2013

Dininnyj S (2012) http://www.businessweek.com/ap/financialnews/D9IQS0C01.htm. Accessed 19 Oct 2012

Federal Energy Regulatory Commission (1996) Final environmental impact statement, condit hydroelectric project, Skamania and Klikitat Counties, Washington. www.pacificorp.com/content/dam/pacificorp/doc/Energy_Sources/Hydro/Hydro_Licensing/Condit/Final_Environmental_Impact_Statement_FEIS_1996.pdf

Kibler KM, Tullos DD (2013) Cumulative biophysical impact of small and large hydro power development in Nu River, China. Water Resour Res J Amer Geophys Union 49(6):3104–3118. doi:10.1002/wrcr.20243

Rashad SM, Ismail MA (2000) Environmental impact assessment of hydro-power in Egypt. Appl Energy 65(1–4):285–302

Chapter 3
Rainmaker in Kuwait: Precipitating a Solution

Pricing the Priceless

*The fact that this installation is now fully operational and
has generated water out air confirms our technology is
equipped to solve real-world problems.*
—Dutch Rainmaker, September 2013

*When I first saw the technology I knew that I was looking at
something that had the ability to dramatically change how
water is supplied to communities around the world.*
—Piers Clark
Commercial Director—Thames Water Utilities
Strategy advisory board member—Dutch Rainmaker

Introduction

The *Green Wall* project in Kuwait was an ambitious 10-year initiative to plant
315,000 trees along 420 km of the arid country's borders. The Kuwaiti environmental agency had commissioned the implementation of a pilot project at Umm Al
Himam ("the project site"). On September 4, 2013, environmental technology
major, *Dutch Rainmaker* (dutchrainmaker.nl) had announced validation of its
installation made on behalf of the Kuwait Environment Public Authority (KEPA).
The AW75 *Dutch Rainmaker* water-out-of-air unit had passed through the concept
and pilot phases and had been successfully deployed in a challenging arid region and
was slated to add "significant value" to the *Green Wall* project (Dutch Rainmaker
2013). Discussions on expanding the air-to-water project had been initiated.

The Rainmaker Technology

The schematic representation of the working of the AW75 and a computer-generated
image of the installation is shown in Fig. 3.1 [Figure credit www.dutchrainmaker.nl
]. A view of the pilot installation is presented in Fig. 3.2.

© Springer India 2015   23
S. Sunderasan, *Cleaner-Energy Investments*,
DOI 10.1007/978-81-322-2062-6_3

Fig. 3.1 Schematic
representation of the working
of the AW75 and a computer-
generated image of the system

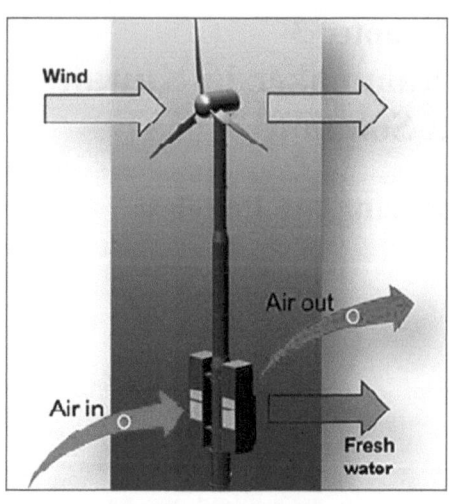

In general, when temperature dropped below the "dew point," water vapor coa-
lesced into droplets of water. The *Rainmaker* system, consisted of stand-alone direct-
drive wind turbines, erected in warm and humid regions, arid and dry regions, or in
regions without water resources and related infrastructure. The mechanical energy of

Fig. 3.2 The AW75 water-from-air pilot installation in Umm Al Himam, Kuwait

the rotating blades produced thermal energy through compressors. The heat was then used to cool down large amounts of ambient air using vents. The AW75 turbine forced ambient air through a heat exchanger, where the flowing air was cooled and moisture carried by it was condensed. This condensate was collected in a water storage compartment and could be used for drinking or irrigation, as required.

The company had also been operating a pilot plant in the Netherlands, at Leeuwarden, successfully producing water from condensing moisture carried in the air. The company believed that the technology would be viable for all locations between the two topographical extremes (Williams 2014).

Dutch Rainmaker believed that the technology was best suited for remote locations since it did not need to be connected to an energy grid, existing water infrastructure, reservoir, or groundwater aquifer. It could generate up to 7,500 l of clean water each day, at the very location where water was most in need, thus avoiding transportation and supply of water. The company also proposed to couple the AW75 with a bottling unit for local distribution of water.

Locations across the Middle East parts of South America, Africa, southeast Asia, and Australia with warm and humid climates and remote locations represented attractive markets for the *Rainmaker* product. Subject to the local geography, the system could secure water for drinking as well as agricultural use. When coupled with a treatment unit, the AW75 could provide safe drinking water for islands, particularly at tourist destinations, avoiding transportation of demineralized and packaged water from the mainland.

Production of water at a particular location depended on the ambient temperatures and relative humidity. According to the company's *Product Sheet* for the Air-to-Water (AW) 75 system, air at 30 °C with a relative humidity of 60 % contained approximately 16 g of water per kilogram of air. Likewise, at 35 °C and 80 % relative humidity, air would contain about 30 g of water per kilogram. Founded in 2007, and funded by ICOS Capital, in 2012, the company was honored with the title "Ones to Watch" (Rainmaker 2012), and in October 2012, the *Dutch Rainmaker* was awarded the first *Enlightenmentz of the Year* Award (The Winner in the Product Category 2014) in recognition of:

quote

- The huge potential impact of the *Dutch Rainmaker* on solving the global water shortage
- The proprietary technology and combination of wind energy and direct-drive heat pump
- The wide geographical area of application, and for
- combining best of Dutch water tradition being co-developed with *Wetsus* world-class water knowledge center

unquote (Rainmaker 2014).

Technology Strengths and Risks

The technology developed by USA, Florida-based water technology company, Aqua Sciences, (http://www.aquasciences.com/) was selected as a leader in addressing key global water challenges by US government officials and independent subject experts in March 2014. The patented technology packaged into a water station comprised a concentrated salt compound that attracted water in the air. Water was then extracted by removing the sodium concentrate. The technology was slated to be simple, off-grid, scalable, and capable of working in low-humidity environments. Similarly, the Skywater 300 Advanced Air-to-Water system developed by the Island Sky Corporation (www.islandsky.com) could produce 300 gallons (or about 1100 lpd) of water from humidity. The system pumped ozone through the water to eliminate bacteria and purify it, making it safe for human consumption.

Israel's *Water-Gen*, included in the list of "The World's Most Innovative Companies, 2014" had built a water-generating unit that turned moisture in the air into drinking water, albeit, using solar PV or other sources of electricity. The system consisted of an air-filtering unit, extracting humidity from the air and effectively dehumidifying it with a heat exchange system. The filtered and treated ready-to-use condensate water filled up tanks at the bottom of the unit (Rosbrow 2014). Subject to ambient temperature and humidity conditions, the unit could produce 250–800 l of potable water a day, consuming about 2 US cents (USD 0.02) worth of electricity per liter (The Hindu 2014).

Australian James J. Reidy had developed the *AirWater* machine, inspired by the dehumidifier, to mimic the natural hydrologic cycle. In addition, the system sterilized each drop of water by exposure to ultraviolet (UV) light. The sterilized water was made to pass through a 1-micron activated carbon water filter to remove possible solid particles, toxic chemicals, volatile organic matter and other contaminants and odors, tastes and colors. The filtered water was exposed to UV light and sterilized again. The 20-liter-per-day (lpd) machines were sold at AUD 1,300 (including taxes) through 5,000 lpd units at AUD 160,000 (including taxes). Machines larger than 50 lpd came with the solar PV option and consequently the 5,000-lpd machine with solar PV power cost AUD 250,000 and low to negligible operating expenses (Gizmag 2014). In comparison, the AW55 model comprising the water making unit and the heat pump technology as the AW75 but without the autonomous energy generator was priced at Euro 75,000.[1] Dutch Rainmaker was therefore required to balance between the pricing of water and the first cost of the system.

Select Input Data to Estimate Project Returns

Parameter	Unit	Value	Remarks
Energy usage per liter	USD	0.02	Based on Israeli Water-Gen data
Capex: AW55 machine (without dedicated power supply)	USD	105,000	http://dutchrainmaker.nl/markets/aw55/
Capex: 5,000-lpd machine (without dedicated power supply)	USD	160,000	Based on Australian *AirWater* machine data
Capex: 5,000-lpd machine (with dedicated power supply)	USD	250,000	Based on Australian *AirWater* machine data, with solar PV power system
Water tariff per 1,000 l proposed for Kuwait	USD	1.00	Milan Milutinovic, Katharine A. Murtaugh, and Elfatih A.B. Eltahir, "A Proposal for Water Pricing in Kuwait", home.etf.rs/~vm/papers/Milan_paper.pdf, last accessed 8 May 2014
Fixed O&M costs	% of capex/a	1.00**	

Annualized returns on the AEX All Share Index Jan 1, 2012—May 8, 2014: 13.95 %**

**Author's estimates/computations

---

[1] http://dutchrainmaker.nl/markets/aw55/, last accessed 8 May 2014.

Teaching Note

Case Synopsis

Dutch Rainmaker of the Netherlands had developed a wind-energy-driven system that would generate energy required to extract the vapor from the flowing air and to condense it into water for routine use. The water accumulated in tanks within the system. Several other vendors had developed and deployed such water-extraction systems, especially for use in emergency situations such as earthquakes and hurricanes. At about 7,500 l per day, the Rainmaker system was larger than most comparable technologies. Additionally, since it was coupled with a wind energy generator, it did not require external power supply: It was totally independent of the utility power grid, surface water sources, groundwater aquifers, and water distribution networks.

The pilot installation in Kuwait/Umm Al Himam was intended to irrigate avenue plantations along the country's borders. However, for the company to make rapid inroads in the Middle East, Australia and other parched regions where it might be possible to extract moisture from the air, the company needed to optimize between the first cost and the tariff on offer for the water so supplied.

Case Question

A key learning outcome of the case is the sensitivity and scenario analyses for a monopoly vendor supplying a product that is vital for sustaining life on the planet and one with no substitutes: water. Analysts would need to arrive at price points for the technology and for the water supplied to ensure reasonable returns to equity investors.

Teaching Objectives

- Enabling students extract case facts, relevant data from multiple sources and to make reasonable assumptions/estimates.
- Preparing cash flow estimates for a stand-alone project.
- Computing returns on the project (before debt service) and for equity holders.
- Undertaking sensitivity and scenario analyses to arrive at a combination of input parameters that would balance between the desired returns to investors and the price paid for the water supplied.

Case Objectives and Use

The primary objective of the case is collecting data and assembling a coherent input data set. The technology was at a nascent stage of development, and two pilot units were installed. Market prices for the system were yet to be determined or made public. Course participants would therefore require to work backward to arrive at a system price (capital cost) that would make the (water supply) project viable by piecing together of the puzzle, eventually taking the shape of the financial model which provides for the development of multiple scenarios.

The plant is assumed to work for 300 days a year, generating 7,500 l each day. In the present case, the capital structure (80 % debt and 20 % equity) is assumed. The average return on the NYSE Euronext all-stock index is employed as a benchmark for returns on equity investments. The tariff per kiloliter of water sold is varied to generate multiple return scenarios for the equity invested.

To keep the analysis simple, exchange-rate-related issues and taxes are ignored. The cost of debt and the debt–equity ratios could also be varied in keeping with the lenders' perception of project risk. At higher costs of project debt, and at assumed levels of leverage, the participants could assess the prospect of the project defaulting on its debt service obligations. Such scenarios would then need to be systematically tabulated and addressed *ex-ante*.

Teaching Plan

The case deals with new technology development and deployment where the vendor/installer could be a monopoly service provider for a product with no substitutes: water. In regions of acute scarcity, the willingness to pay for water could be rather substantial. Hence, water and other such tariffs are regulated by statutory authorities. The analysis herein demonstrates that at the expected level of capital costs and proposed water tariffs of USD 1 per kiloliter, the project would be unviable, despite the fact that 80 % of the project is presumed to be debt funded at a mere 2 % annual rate on declining balance. Hence, the authorities need to raise tariffs, and the developers need to bring down capital costs of the project asset, or both. In fact, at this tariff and this proportion of debt in the capital structure, the project merely manages to earn its costs back over 20 years if the capital costs were about USD 15,000 or about 10 % of the present cost assumed.

Water being fundamental to sustaining life on the planet, having control on the tap provides the vendor with immense market power. However, such power cannot be translated into higher tariffs owing to statutory ceilings on price and more importantly to the threat potential entry by competitors. In the present instance, the pilot had been installed to help irrigate avenue plantation along the country's borders. However, the project could be expanded to serve to provide drinking water for people, their livestock, and plantations, where a price would have to be placed on the water supplied. If the company could reduce capital costs by 10 %, i.e., from USD 150,000 to USD 135,000 and if the water tariffs were to be raised to USD 5.50 per kiloliter, then the project would return 6.48 % and while returns of about 14.14 % would be earned on invested equity: levels comparable to the returns on the all-stock index on the Amsterdam Exchange.

The instructor could guide the students to evaluate the project in a number of small steps as laid out herein below:

1. Table 1—Input parameters based on case facts and author's assumptions/ estimates
2. Table 2—EBITDA margins on project
3. Table 3—Loan amortization schedule
4. Table 4—Sensitivity and scenario analyses for various levels of capital expenditure and water tariffs.

The cash flowchart for the project would need to be drawn up one step at a time:

1. Input data for base case

Parameter/Description	Unit	Value	Remarks
Project cost			
Condensation system cost	USD	105,000.00	AW55 pricing without the power supply unit
Wind turbine component	USD	45,000.00	2 US cent per liter; author' est.
Total capital expenditure		*150,000.00*	
Means of finance			
Sponsor equity	20 %	30,000	Author's est.
Project debt	80 %	120,000	Author's est.
Water output per day	liter	7,500.00	Case fact
Working days in a year	days	300.00	Author's est.
Water output per year	liter	2,250,000	Author's est.
Price of water	USD per liter	0.001	Proposed
Operating expenses	of Capex	1.00 %	Author's est.
System life	year	20	Author's est.
Interest rate	per annum	2.00 %	Author's est.

2. Margins earned on project (before debt service): shown for 8 of 20 years

Year		1	2	3	4	5	6	7	8
Capital expenditure	USD	150,000							
Annual water production	liter	2,250,000	2,250,000	2,250,000	2,250,000	2,250,000	2,250,000	2,250,000	2,250,000
Water tariffs	USD	0.001	0.001	0.001	0.001	0.001	0.001	0.001	0.001
Project revenues		2,250	2,250	2,250	2,250	2,250	2,250	2,250	2,250.00
Operation and maintenance	USD	1,500	1,500	1,500	1,500	1,500	1,500	1,500	1,500.00
EBITDA		750.00	750.00	750.00	750.00	750.00	750.00	750.00	750.00

3. Loan amortization schedule: shown for 8 of 20 years

	1	2	3	4	5	6	7	8
Principle at the beginning of the year	120,000	114,000	108,000	102,000	96,000	90,000	84,000	78,000
Principle repaid during the year	6,000	6,000	6,000	6,000	6,000	6,000	6,000	6,000
Principle outstanding at the end of the year	114,000	108,000	102,000	96,000	90,000	84,000	78,000	72,000
Average principle outstanding during the year	117,000	111,000	105,000	99,000	93,000	87,000	81,000	75,000
Interest charge	2,340	2,220	2,100	1,980	1,860	1,740	1,620	1,500

4. Sensitivity of project returns (IRR) to capital expenditure

Capex (USD)	Project IRR (%)
14,904	0
13,449	1.06
12,192	2.06
11,680	2.51
11,144	3.01
10,647	3.51
10,181	4.01
9,298	5.06

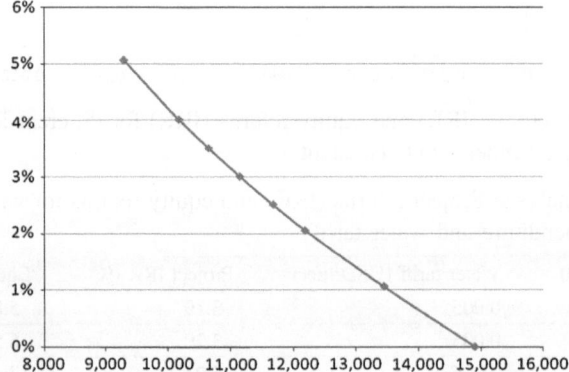

Graph Project returns (IRR) for different levels of capital expenditure, holding all other inputs constant.

5. Sensitivity of Project returns (IRR) and equity returns to water tariff

Water tariff (USD/liter)	Project IRR (%)	Equity IRR (%)
0.01	12.72	74.63
0.009	10.93	54.85
0.008	9.06	39.22
0.007	7.08	26.49
0.006	4.96	15.62
0.005	2.64	5.31

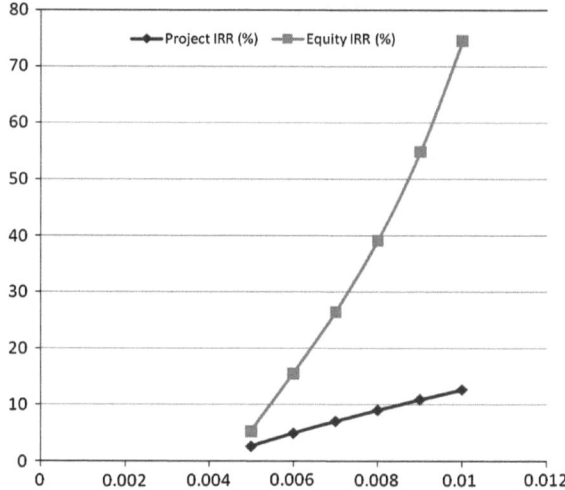

Graph Project returns (IRR) and equity returns (IRR) for different levels of water tariffs, holding all other inputs constant.

6. Scenario analysis: Project returns (IRR) and equity returns for various levels of capital expenditure and water tariff

First cost (USD)	Water tariff (USD/liter)	Project IRR (%)	Equity IRR (%)
142,500	0.005	3.19	5.89
135,000	0.005	3.79	6.52
135,000	0.006	6.24	17.8
127,500	0.006	6.96	19.08
127,500	0.007	9.28	32.46
120,000	0.007	10.16	35.07

References

Gizmag (2014) Extracting Water From the Air. http://www.gizmag.com/extracting-water-from-the-air/2796/picture/87/. Accessed 8 May 2014

Rainmaker D (2012) Cleantech Connect 'Ones To Watch'. http://www.cleantechinvestor.com/portal/the-ones-to-watch2012/11231-dutch-rainmaker.html. Accessed 8 May 2014

Rainmaker D (2013) Dutch Rainmaker's Pilot Installation Successfully Generates Water Out of air in Umm Al Himam, Kuwait. http://www.dutchwatersector.com/news-events/news/7399-dutch-rainmaker-s-pilot-installation-succesfully-generates-water-out-of-air-in-umm-al-himam-kuwait.html. Accessed 5 May 2014

Rainmaker D (2014) Dutch Rainmaker Awarded First Enlightenment of the Year Award. http://dutchrainmaker.nl/news/press/. Accessed 8 May 2014

Rosbrow L (2014) In major breakthrough, an Israeli Company has created water out of thin air. http://www.policymic.com/articles/85195/in-major-breakthrough-an-israeli-company-has-created-water-out-of-thin-air. Accessed 8 May 2014

The Hindu (2014) Now, A Machine That Makes Drinking Water From Thin Air. The Hindu. http://www.thehindu.com/news/international/world/now-a-machine-that-makes-drinking-water-from-thin-air/article5956268.ece. Accessed 8 May 2014

The Winner in the Product Category (2014) http://www.kennislink.nl/publicaties/enlightenmentz-of-the-year. Accessed 8 May 2014

Williams A (2014) Wind of change: water from air. Water Wastewater Int, p 28–31

Chapter 4
Amakhala Wind: Turbulence in the Detail?

Optimizing Project Capital Structure

> We believe these are initial steps that can open out other opportunities.
>
> —S. Ramakrishnan,
> Executive Director (Finance),
> TATA Power Limited,
> June 2013

> Everything we do and deliver today will allow others to realize their vision tomorrow.
>
> The Exxaro brand promise

Background

The proposed 650 MW in capacity additions including two hydro-electricity projects in Georgia and wind energy generation in South Africa represented a strategic diversification of TATA Power's portfolio away from its home market in India, where it had faced adverse market conditions in the immediate past.[1] The projects were slated for commissioning in 2016 and the company was to invest between USD 150 and 200 million over a three year horizon.

The Project Company

TATA Power (www.tatapower.com) was India's largest integrated power company with a trans-continental footprint, with its value chain stretching from fuel extraction and logistics to power generation (from coal, wind, solar, hydro, geo-thermal, etc.), transmission, trading, and distribution. Cennergi (Pty) Limited

---

[1] Katya B. Naidu, "Tata Power's Nimble Steps See Lower Equity Investment," *Business Standard*, 23 June 2013.

© Springer India 2015
S. Sunderasan, *Cleaner-Energy Investments*,
DOI 10.1007/978-81-322-2062-6_4

(www.cennergi.com), a 50:50 joint venture between TATA Power Limited and Exxaro Resources Limited (www.exxaro.com), was incorporated in March 2012 to develop, build, and operate the 138.6 MW Amakhala Wind Farm project among others. The joint venture partner, commodity, and mining major Exxaro (www.exxaro.com) were South Africa's second-largest coal producer and the world's third-largest producer of mineral sands. In addition, the company had an exposure in iron ore, base metals, and ferroalloys. *Cennergi* was also slated to continually explore possibilities for expanding and diversifying its portfolio with such other projects involving wind energy, coal, gas, and oil-fired electricity generation in South Africa, Botswana, and Mozambique. The shareholding structures and the spread of *Cennergi's* South Africa projects are depicted in Fig. 4.1.

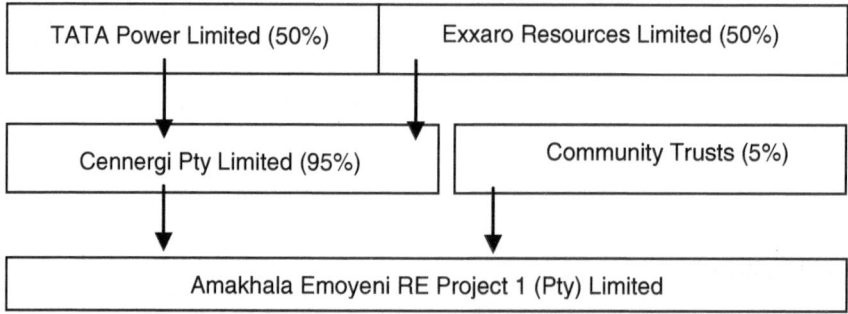

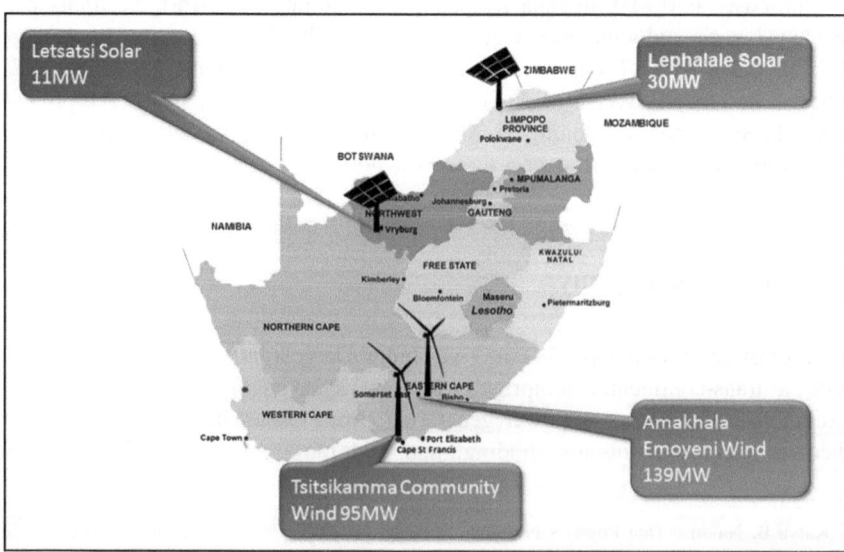

Fig. 4.1 Project Company Shareholding Structure, and South Africa Projects developed by Cennergi (Pty) Limited (www.cennergi.com/programmes, last accessed 9 July 2013)

Project Description

The 138.6 MW green-field Amakhala wind energy project was to be undertaken through a special purpose vehicle (SPV), the Amakhala Emoyeni RE Project 1 (Pty) Limited.[2] The SPV ("Amakhala") would sell the generated power to the South African electricity utility, Eskom (www.escom.co.za) under a 20 year power purchase agreement (PPA) (Amakhala Wind 2013). The organization structure and financial statements of Eskom are annexed to this case document.

The project awarded under the South African Renewable Energy Independent Power Producer's Procurement Programme (REIPPPP), was located near the town of Bedford in the Eastern Cape. In addition to the wind resource, the site was supported by the grid infrastructure to evacuate the power generated. The project was sub-grouped into seven commercial wind farms and *Cennergi* had executed long-term lease agreements with the landowners concerned. The project was designed to minimize adverse impacts on bird and bat life, while 98 % of the leased land would still be available as pasture land for grazing.[3]

The Amakhala project had achieved financial closure by end-May 2013, construction was to commence early in calendar 2014 and the wind farms were expected to be operational by the third-quarter of year 2016. A detailed environmental management plan had been compiled to ensure social and environmental sustainability during the construction, operation, and decommissioning phases of the project.

Germany/Hamburg based Nordex had been contracted to supply the 134.4 MW capacity farm with 56 of its N117/2,400 turbines (2.4 MW each installed at 80 m hub height) and to provide service support for a period of ten years. In a press release dated June 14, 2013, the company announced that construction work would commence in Q3—2014, installation would follow starting Q2—2015 and that the wind farm would be handed over for operations in June 2016[4] (Louise Downing 2013).

The International Finance Corporation, private sector lending arm of the World Bank Group (IFC) had approved project debt on the grounds that the initiative would:

---

[2] The South African Independent Power Production program required local communities to hold a 5 % equity stake projects: The Amakhala Emoyeni Cookhouse Wind Farm Community Trust and the Amakhala Emoyeni Bedford Wind Farm Community Trust would each hold 2.5 % stake in the SPV. The trusts do not inject equity capital but receive dividends that are applied to local development and community projects. In addition, the SPC had committed to spend 1.5 % of revenues on socioeconomic development and 0.6 % of revenue on enterprise development in the region.

[3] http://www.cennergi.com/programmes/wind-power/, last accessed 9 July 2013.

[4] http://www.nordex-online.com/index.php?id=53&L=2&tx_ttnews[tt_news]=2404&tx_ttnews [backPid]=45&cHash=e3a64ec086, last accessed 9 July 2013.

Quote:

(i) promote additional employment and skills development during construction and operation;
(ii) diversify South Africa's energy mix;
(iii) reduce carbon emissions by producing green, emission-free electricity;
(iv) contribute to increased generation capacity in South Africa (as part of the planned system expansion);
(v) contribute to the upliftment of local communities, through financial benefits flowing from a 5 % carried shareholding to be held via two local community trusts; and
(vi) provide positive demonstration effects in South Africa and the neighboring countries, as one of the initial large wind projects in the region.

Unquote.[5]

Project Cost and Means of Finance

The total project cost was estimated at ZAR 3.945 billion ($\sim$ USD 395 million) of which, about 70 % was slated for the purchase of the wind-turbine hardware. The ZAR denominated project debt of ZAR 3.156 billion, pegged at 4 times the equity component, was mobilized from the IFC, member World Bank Group, (ZAR 700 million), equivalent to USD 70.7 million[6] and the remaining from the Standard Bank Group (Amogelang Mbatha 2013).[7] *Cennergi* contributed the equity investment amounting to ZAR 789 million.[8] Among other things, project analysts were required to estimate the cost of project debt which was not made public by the parties involved.

Project Strengths and Risks

The South African competitive bidding process for RE projects required bidders to have fully developed projects including land rights, water rights, access to power transmission, commitments for financing, and one year's worth of meteorological station data for the proposed site. The collection of on-site data ensured that project

---

[5] See footnote 3 above.

[6] 1 USD = 10.1992 on 9 July 2013, http://www.oanda.com/currency/converter/ (rounded off to 1 USD = 10 ZAR for estimations).

[7] Funding Pours In for Tata Power JV, 5 June 2013, http://www.gtreview.com/trade-finance/global-trade-review-news/2013/June/Funding-pours-in-for-Tata-Power-JV_10948.shtml, last accessed 9 July 2013.

[8] http://www.tata.com/company/Media/inside.aspx?artid=pbELtVk3o2M= Business Standard, 5 June 2013, last accessed 9 July 2013.

expectations were more realistic and minimized risks of overestimation of resource availability and hence of project revenues. Competition, therefore, was among serious and qualified players with real projects. Further, payments against supply of power to Eskom were guaranteed by the South African government (Susan Kraemer 2013).

The average "levellized" tariff for wind energy during the "second window" of bidding being discussed herein was ZAR 0.89/ kWh (South African Wind Energy Association 2013). A ZAR 0.9114 tariff per kWh was embedded within the computations submitted by the company to the CDM Executive Board in December 2011. *Cennergi* CEO, Thomas Garner, was quoted confirming the attractiveness of power sector investments in South Africa owing to policy certainty and on possible dollar returns on projects ranging between 12 and 13 % (Sue Blaine 2013). The PPA signed under the program was a "take-or-pay" contract, i.e., Eskom had committed to pay the tariff on the generated power, irrespective of demand. Further, the tariff itself was indexed to the rate of inflation, and was to be escalated accordingly, over the duration of the contract (Paul Semple 2013). Project risks included the exposure to volatility in the ZAR/USD exchange rates, uncertainty with the sustained availability of the utility grid to evacuate the generated power, and an increase in the capital cost of the turbines, ancillary equipment and spares and natural force majeure situations. Political risks included mandatory nationalization of project assets, or limitations on repatriation of profits out of the country at some time in the future.

Select Input Data To Estimate Project Returns

Extract from Project Design Document Submitted to the CDM Executive Board, Project Reference 7576; December 21, 2011, p. 16. http://cdm.unfccc.int/Projects/DB/CarbonCheck_Cert1349249257.11/view

Parameter	Unit	Value	Remarks
Load factor of the wind farm	Ratio	0.305	Based on field assessments; 0.27 as per RE tariff guideline for wind farm projects. Annual electricity production: capacity (MW) × PLF (%) × 8,760 (hr)
The period of assessment	Years	20	Guidelines on the assessment of investment analysis; PPA tenure
Electricity tariff[a]	ZAR/ kWh	0.6585	Media statement "NERSA's decision on Eskom's required revenue application—multi-year price determination 2010/11−2012/13 (MYPD 2)" February 24, 2010, page 2, paragraph 1: www.eskom.co.za/content/MediaStatementMYPD2 ∼ 1.pdf
Fixed O&M costs	USD/ kW/a	24	
Variable O&M costs	USD/ kWh	0	

Annualized returns on the Johannesburg all share index June 2008−June 2013: 11.23 %[b]

[a] Average tariff for REIPPP bidding Round 2 was ZAR 0.89 or about US cent 8.9/kWh
[b] Author's computations

Teaching Note

Case Synopsis

Energy generation from intermittent sources such as the sun and wind is faced with inherent resource uncertainty. However, project developers build such uncertainty within their projections and develop their business plans around such assumptions. In most situations, challenges relating to regulatory certainty, optimal pricing, connectivity to the grid, and off-take of the generated electric power are more daunting than the natural factors. The case outlines a situation where South Africa, a resource abundant, power deficient economy, had announced an ambitious program to enhance power generation capacity using cleaner energy technology including solar photovoltaics, concentrating solar power, and wind turbines.

Cennergi, a joint venture between Tata Power of India and Exxaro Resources of South Africa had proposed the installation and operation, of power generation units across technologies, spread over southern Africa commencing with the 134.4 MW Amakhala Wind Power project in the Eastern Cape province of South Africa. The wind-turbine equipment was ordered on Nordex of Germany and the farm was to come on stream by mid-2016. The company had applied for registration with the UNFCCC's CDM Executive Board for the approval and issue of emission reduction credits. The power generated was to be sold to Eskom, the South African utility under a "pay-or-take" contract.

Case Question

A key learning outcome of the case is the collection and collation of data from disparate sources to develop a financial model, to validate assumptions and to assess the viability of the project under various scenarios.

Teaching Objectives

- Enabling students extract relevant data from multiple sources to prepare cash flow estimates for a standalone project.
- Comparing data from various sources and arriving at the most plausible combination of inputs.
- Preparing a financial model for the project.
- Making a list of (unavailable) input data that need to be assessed or computed indirectly from projected returns or other eventual outputs from the financial model.
- Taking cognizance of various quantitative inputs and subjective evaluations of risk and returns and making recommendations on the viability of the proposed ventures.

Case Objectives and Use

The primary objective of the case is collecting relevant data from multiple sources and assembling a coherent data set. This is necessitated by the unavailability of key data in the public domain owing to the confidentiality of several proposals and agreements. Course participants would do well to observe the piecing together of the puzzle, eventually taking the shape of the financial model which provides for the development of multiple scenarios.

In the present case, the capital structure (80 % debt and 20 % equity) is revealed. The expected return on investment is 12−13 % or about 12.5 % on average. The tariff per kWh of electric power generated is within the range ZAR 0.89 and ZAR 0.9137. The plant load factor is estimated between 27 and 30 % for the project. Each of these figures is extracted from direct or indirect mention across submissions and reports.

Local currency (ZAR) debt is mobilized from the IFC and from Standard Bank. The cost of debt is not known and needs to be computed working back from project returns. Further, the participants would need to assess the prospect of the project running illiquid by opting for a high degree of leverage. The conditions that could lead to such illiquidity need to be laid out and addressed ex-ante.

Teaching Plan

The case presents a project launch scenario where a preferred bidder needs to balance between financial returns and socioeconomic and developmental returns while also staying ahead of the competition. Cennergi had made the cut as a preferred bidder on the basis of a strong proposal and on the strength of the promoters' record in energy service delivery. Cennergi had ordered wind-turbine equipment on Nordex of Germany. Investors generally assumed project risks including construction and commissioning risks, while lenders essentially absorbed the risks of uncertain and delayed cash flows generated by the project. The lenders essentially took a call on the off-taker's ability and willingness to pay for the power procured. It is in this context that the strength and payment record of Eskom was key, further backed by a sovereign guarantee. In this sense, the project could be seen as being of low payment risk and almost devoid of exchange rate risks. The instructor would do well to get the course participants to analyze the financial strengths of Eskom and the merits of a government guarantee backing up the "take-or-pay" contract. The instructor could guide the discussion on enhancing project profitability by reducing upfront costs, enhancing revenues, or both.

The instructor could guide the students to evaluate the project in a number of small steps as laid out herein below:

1. Table 1—Replicating the net cash flow model presented to the CDM Executive
   Board.
2. Table 2—Adjusting the net cash flow model presented for more recent project
   cost estimates and other input data.
3. Table 3—Estimating the capital cost of the project if the equity investors sought
   a return of 12.50 %
4. Table 4—Computing the return on equity if the project debt were to be interest-
   free, in the manner of evaluating an extreme case.

What Happened Next

As of the 14th of July, 2013, the project was as yet awaiting registration with the
CDM—Executive Board for emission reduction certificates equivalent to mitigating
370,665 metric tonnes of CO_2 equivalent per annum[9] in response to the request
submitted in December 2012. Savannah Environmental Private Limited had filed
for an amendment requesting that the 140 MW Phase 2 of the project (DEA
Reference No: 12/12/20/1754/2) be renamed "Mesenge Emoyeni," since it was
being developed by different shareholders and developers.[10]

The cash flow chart for the project would need to be drawn-up one step at a time:

1. Input data/replication of base case

Capacity (MW)	140
PLF (efficiency factor) (%)	0.305
Hours per year (no)	8,760
Model tariff (USD/kWh)	0.091137
Fixed O&M (USD/kW/annum)	24

[9] http://cdm.unfccc.int/Projects/DB/CarbonCheck_Cert1349249257.11/view.

[10] Savannah Environmental Pty Ltd. "Motivation for Amendment of Environmental Authorization,"
January   2013,   http://www.savannahsa.com/documents/4715/Amakhala%20Emoyeni%20Phase%
202%20Wind%20Energy%20Facility%20and%20Associated%20Infrastructure,%20Eastern%20Cape
%20Amendment.pdf. Last accessed 14 July 2013.

	Unit	2012	2013	2014	2015	2016	Contd.	2034	2035
Investment cost	USD '000	−10523.3	−126280	−126,280	−52616.7				
Annual electricity supply to eskom	MWh				187,026	374,052		374,052	374,052
Income from electricity sale	USD '000				17044.99	34089.98		34089.98	17044.99
Cost of electricity generation	USD '000				3,360	3,360		3,360	3,360
Total income from project implementation	USD '000				13684.99	30729.98		30729.98	13684.99
Net cash flow	USD '000	−10523.3	−126280	−126280	−38,931.7	30729.98		30729.98	13684.99
Project IRR		6.6463 %							

Reproduced as presented in the CDM—Project Design Document (PDD)

2.   Revised project inputs/assumptions and project IRR

Description	Unit	Proportion	Input value
Project cost			
Wind site	ZAR	30 %	1,183,500,000
Wind-turbine equipment	ZAR	70 %	2,761,500,000
Total	ZAR		3,945,000,000
Means of finance			
Equity	ZAR	20 %	789,000,000
Debt	ZAR	80 %	3,156,000,000
Total			3,945,000,000
Terms of finance			
Investor expectation	%		12.50 %
Rate of interest	%		?
Tenure of the loan	year		18

Depreciation

		Year			20		
Useful life of the asset		Year			20		
Straight-line method (SLM)		%			5.00 %		
Taxation		%			28.00 %		
Year				1	2	Contd.	21
Revenue							
Revenue from the sale of power				170885656	341771312		170885656
Sale of assets							
Total revenue				170885656	341771312		170885656
O&M expenses				16,800,000	16,800,000		8,400,000
Total expenditure				16,800,000	16,800,000		8,400,000
(In proportions as projected by the company)		3.33 %	40 %	40 %	16.67 %		
Cash outflow	−3,945,000,000	−131,368,500	−1,578,000,000	−1,578,000,000	−657,631,500		
EBITDA				154,085,656	324,971,312		1,624,85,656
Net cash flow from project (before tax)		−131,368,500	−1,578,000,000	−1,578,000,000	−503,545,844	324,971,312	162,485,656
Project IRR	**4.8361 %**						

Source Author's computations

3. Revised project cost if project IRR were to be 12.50 % (rounded), other parameters held constant

Year		1	2	Contd.	21
Revenue					
Revenue from the sale of power		*170,885,656*	*341,771,312*		*170,885,656*
Sale of assets					
Total revenue		170,885,656	341,771,312		170,885,656
O&M expenses		16,800,000	16,800,000		8,400,000
Total expenditure		*16,800,000*	*16,800,000*		*8,400,000*
(In proportions as projected by the company)	3.33 %	40 %	40 %		16.67 %
Cash outflow	**-2134898286**	-71,092,112.94	-853,959,314.6	-853,959,314.6	-355,887,544.3
EBITDA		154,085,656	324,971,312		162,485,656
Net cash flow from project (Before Tax)	-71,092,113	-853,959,315	-853,959,315	-201,801,888	324,971,312	162,485,656
Project IRR	**12.4612 %**				

Source Author's computations

4. Cash flow Statement computing IRR at zero interest (0.02 %), straight-line depreciation of plant and machinery; other parameters held constant

Year					1	2	3	21
Revenue								
Revenue from the sale of power					170,885,656	341,771,312	341,771,312	170,885,656
Sale of assets								
Total revenue					170,885,656	341,771,312	341,771,312	170,885,656
O&M Expenses					16,800,000	16,800,000	16,800,000	8,400,000
Total expenditure					16,800,000	16,800,000	16,800,000	8,400,000
(In proportions as projected by the company)		3.33 %	40 %	40 %	16.67 %			
Cash outflow	-3,945,000,000	-131,368,500	-1,578,000,000	-1,578,000,000	-657,631,500			
EBITDA					154,085,656	324,971,312	324,971,312	162,485,656
Net cash flow from project (before tax)		-131,368,500			-503,545,844	324,971,312	324,971,312	162,485,656
Project IRR	**4.8361 %**							
Interest payment					707,095	677,218	637,382	0
Depreciation					138,075,000	138,075,000	138,075,000	138,075,000
Earnings before Tax					15,303,561	186,219,094	186,258,931	24,410,656
Taxation					4,284,997.01	52141346.39	52152500.57	6834983.74
Earnings after Tax					11,018,564	134,077,748	134,106,430	17,575,672
Principal repayment					-87,666,667	-175,333,333	-175,333,333	0
Depreciation added back					138,075,000	138,075,000	138,075,000	138,075,000
(In proportions as projected by the company)		3.33 %	40 %	40 %	16.67 %			
Investment cash in flow		-26,273,700	-315,600,000	-315,600,000	-131,526,300			
Cash flow from operations					61,426,897	96,819,415	96,848,097	155,650,672
Net cash flow available for investors					-70,099,403	96,819,415	96,848,097	155,650,672
Equity IRR	**10.60377 %**							

Source Author's computations

Appendix: Organization Structure of Eskom Holdings SOC Limited (and Major Subsidiaries)

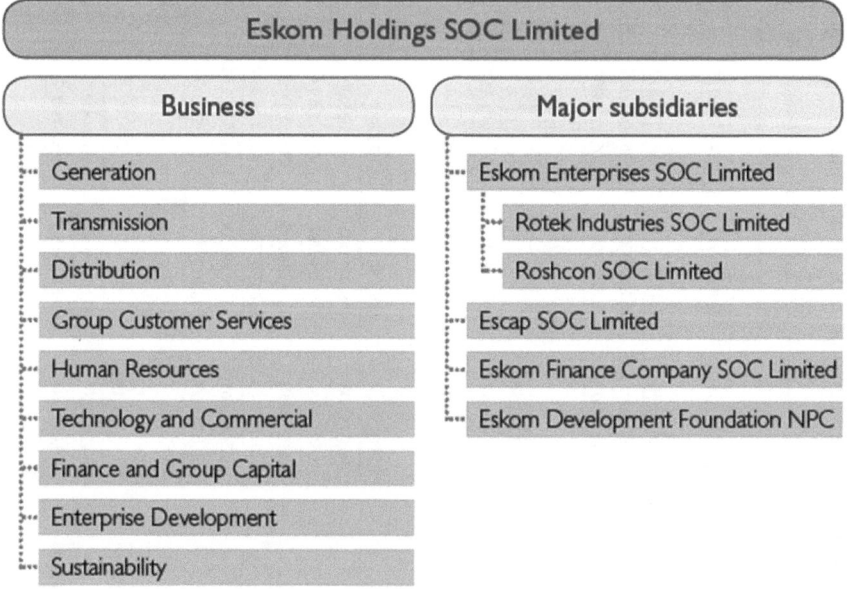

Source http://www.eskom.co.za/c/article/583/organisational-structure/; July 9, 2013.

References

Amakhala Wind (2013) Summary of Proposed Investment, Public disclosure document posted by the International Finance Corporation (IFC), member World Bank Group. 8 Feb 2013. http://ifcext.ifc.org/ifcext/spiwebsite1.nsf/DocsByUNIDForPrint/829882578D9E6A0685257B0C00631A3E?opendocument. Accessed 9 July 2013

Amogelang Mbatha (2013) Exxaro, Tata Secure 7 Billion Rand for Two S. African Wind Farms, Bloomberg.com, 12 June 2013. http://www.bloomberg.com/news/print/2013-06-12/exxaro-tata-secure-7-billion-rand-for-two-s-africa-wind-farms.html. Accessed 9 July 2013

Downing L (2013) Nordex Wins Order From Tata Venture for South African Wind Farm, Bloomberg.com, 17 June 2013. http://www.bloomberg.com/news/2013-06-17/nordex-wins-order-from-tata-venture-for-south-african-wind-farm.html, Accessed 9 July 2013

Paul Semple (2013) SA is in Good Stead In Clean Energy Race, The Sunday Independent 26 May 2013. http://www.sapressonline.com/allenzimbler/wp-content/uploads/2013/05/Sunday-Independent-26-May-13.SA-is-in-good-stead-in-clean-energy-race.htm. Accessed 10 July 2013

South African Wind Energy Association (2013) http://www.sawea.org.za/index.php?option=com_content&view=article&id=120:the-low-cost-of-wind-power-and-how-it-softens-the-eskom-price-impact&catid=11:blog&Itemid=74#. Accessed 10 July 2013

Sue Blaine (2013) Wind, Solar Power Plants Being Built, Thanks to Policy Certainty, Business Day. 20 February 2013. http://www.bdlive.co.za/business/energy/2013/02/20/wind-solar-power-plants-being-built-thanks-to-policy-certainty. Accessed 10 July 2013

Susan Kraemer (2013) South Africa PV: Path to a Unique Bidder Market, PV Insider. 9 July 2013. http://news.pv-insider.com/photovoltaics/south-africa-pv-path-unique-bidder-market?utm_source=http%3a%2f%2fuk.pv-insider.com%2ffc_csp_pvlz%2f&utm_medium=email&utm_campaign=PV+ebrief+9+Jul+13+en&utm_term=South+Africa+PV%3a+path+to+a+unique+bidder+market&utm_content=90793. Accessed 10 July 2013

Chapter 5
San Cristobal Wind Power Project: Addressing Petrel and Diesel Conservation

Making Partnerships Work

The e8 team approached this work with a level of caution akin to the curators responsible for da Vinci's Mona Lisa or Michelangelo's David.
—Michael G. Morris, CEO, American Electric Power (AEP)

Introduction

Made famous by Charles Darwin through his analysis of finches, the Galapagos Islands were declared a World Heritage site by the UNESCO in 1978 and hence the conservation of natural resources and the protection of biological diversity acquired special relevance. A large proportion of the electricity on the islands was generated using imported diesel and transported from continental Ecuador in small tankers, requiring frequent deliveries, given the small storage capacity on the islands.

The eastern most of the Galapagos Islands as shown in Fig. 5.1, San Cristobal, an island of about 6,000 residents, was especially vulnerable to oil spills during fuel transportation. Eighty-five percent of the island was part of the Galapagos National Park. The remaining 15 % was primarily agricultural land and the project was located on highlands outside the National Park area. The San Cristobal Wind Power Project was a partnership among the Government of Ecuador, the United Nations Foundation and a specially constituted Wind Power Commercial Trust instituted by the e7 (later e8). e7, (www.e7.org, later renamed e8) was a coalition of nine electricity supply companies from the G7 countries (later G8).[1] The e7 fund for Sustainable Energy Development was a nonprofit institution recognized as a non-governmental organization (NGO) with special consultative status, by the United Nations Economic and Social Council.

However, the power generation project was as much about the critically endangered Galapagos Petrel (shown in Fig. 5.2) that nested on the island, as it was

---

[1] USA, France, Italy, Canada, Japan, Germany, and the United Kingdom (G7) with Russia (G8).

© Springer India 2015
S. Sunderasan, *Cleaner-Energy Investments*,
DOI 10.1007/978-81-322-2062-6_5

Fig. 5.1 Location of the
Galapagos Islands and San
Cristobal Island

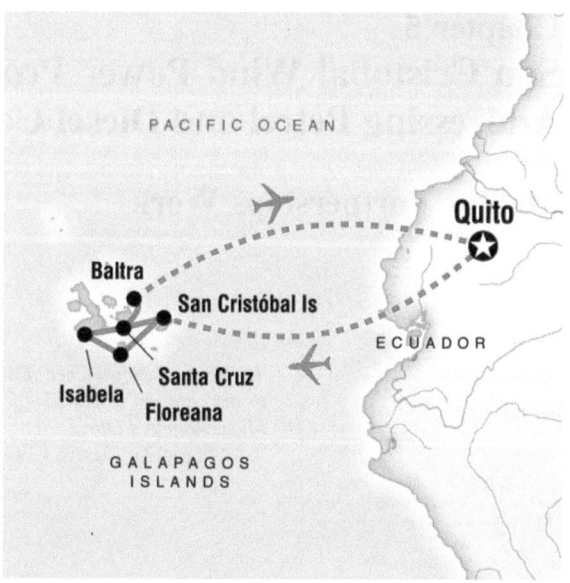

Fig. 5.2 The Galapagos
Petrel—native, nesting,
endangered. *Figure credit*
galapagoswind.org;
reproduced with permission

about the wind energy generators themselves. In January 2001, fuel tanker *Jessica*
struck a reef and broke and spilt an estimated 75,000 gallons of fuel oil and
70,000 gallons of diesel affecting the marine flora and fauna. Much of the marine
life was restored through prompt action by the large number of recovery teams
pressed into the operation.

The incident, however, motivated the international community to explore
alternatives to diesel with a view to preserving the rich diversity of land and marine
life on the islands, including iguanas, sea lions, giant tortoises, penguins, flightless
cormorants, and colorful tropical birds.

The San Cristobal Wind Project was the first stage of an integrated program
supported by Ecuador and the United Nations Development Program (UNDP) that

would eventually bring renewable electricity—hybrid wind-diesel and solar photovoltaic—to the 30,000 residents of the Galapagos archipelago's five inhabited islands.

The project involved superseding 3 × 650 kW of diesel-electric generation capacity with three wind energy generators, each of 800 kW capacity, located on the south face of an old volcanic ridge. The 2,400 kW of wind energy generation capacity, though operating in conjunction with the existing diesel generation capacity, was to initially supply 50 % of the island's annual electricity demand. The power was to be evacuated through the existing grid of the Galapagos Electric utility, *Elecgalapagos*, whose personnel were to be trained in the operation and maintenance of the turbines.

The site was available for cattle and horses to continue to graze freely even after wind farm installation. Notwithstanding the outstanding environmental benefits slated to be delivered, the most compelling reasons for project implementation were ultimately economic in nature. The cost of power generation using diesel was of the order of $0.16 per kWh while the consumer tariff was approximately $0.1 per kWh with the difference borne by the government. The National Council of Electricity had signed a 12-year power purchase agreement (PPA) with the implementing agency, the Trust, at a tariff of $0.1282, lowering the government's subsidy burden by about half. More significantly, frequent fuel spills and tankers running aground had generated bad publicity for the island, adversely impacting tourist arrivals, and hence affecting livelihoods.[2]

The project enjoyed worldwide visibility and extensive international media coverage which puts severe pressure on the implementing agencies and thus, compounded the risks relating to timely project implementation. The islanders had no experience or technological wherewithal relating to the installation, operation, or maintenance of wind energy generators. The residents possessed no experience with logistics of this nature including unloading, transporting, or erecting equipment of comparable proportions. MADE, the Spanish equipment supplier was also contracted to build indigenous capacity in operating and maintaining the machines.

Long-period wind resource data was unavailable and the projections involved substantial averaging of data across time periods.

Electricity consumption on the island of San Cristobal in the year 2005 was of the order of 6.55 million kWh generated using 0.542 million gallons of diesel. The wind generators were slated to deliver close to 25 million kWh of electric power to the grid during the first 7 years (2007–2013).

The 2.4 MW of installed wind energy generation capacity was projected to cost $9,952,790. Consortium leader AEP and other e8 partners viewed the time and money spent as an investment eventually leading to commercial projects within host countries such as Ecuador. Table 5.1 provides a listing of the investments by funding source.

---

[2] Project Design Document, e7 Galapagos/San Cristobal Wind Power Project, UNFCCC Reference No. 1255, 25 October 2007, p. 4.

Table 5.1 Project funding by source

Funding source	Based in	Funding amount ($)
e7 members[a]	Canada	2,000
e7/Global 3e	USA	4,848,000
RWE Power AG	Germany	625,640
United Nations Foundation (UNF)	USA	931,988
Municipality of San Cristobal	Ecuador	239,643
Government of Ecuador—FERUM Subsidy (2005)	Ecuador	1,277,604
Government of Ecuador—FERUM Subsidy (2006)	Ecuador	2,027,915
Total		**9,952,790**

Source CDM Project Ref. 1255—annex 2 to project design document
[a] In addition, the e7 partnership's members provided *pro bono* management and technical expertise to support project development (e7, later e8, http://www.globalelectricity.org/galapagos/, last accessed 9 December 2011)

- Transporting equipment to the remote island location added to project costs; in addition, the project was to build a batching plant for concrete and a pier for unloading equipment on the island.
- An access road was built to reach the south ridge of the El Tropezon hill on which the wind farm was located.
- The project cost involved the construction of a 12.0-km transmission line to convey the energy from the wind park to the interconnection point with the utility grid—including a 3.0-km section built below ground—to prevent disruptions to the movement of endangered birds and to their nesting sites.
- The project proposed to fund and implement a protection program for the Galapagos Petrel, an endangered bird, and one of the six endemic marine birds of the archipelago, which nested on the San Cristobal Island.[3]
- As a part of project implementation, a project-specific Web site—www.galapagoswind.org—was launched to keep the community and all other stakeholders updated.
- MADE engineers in Spain were able to monitor and operate the project equipment through secure Internet connections and obtain data for maintenance and troubleshooting.

The e7 Galapagos wind power project was commissioned in September 2007 and was registered with the UNFCCC on May 13, 2008, making it eligible to receive emission reduction credits. The projected power output from the project is listed in Table 5.2. A total of 10,163 CERs had been issued over the four-year period May 2008–May 2012, with the power generation and diesel displacement details as shown in Table 5.3. By October 2013, the wind farm had completed six

---

[3] *ib. id* p. 13.

Table 5.2 Projected electricity supply from proposed project/96.5 % machine availability

Year	Electricity supplied to the grid (kWh/year)
2007	3,316,759
2008	3,387,240
2009	3,471,068
2010	3,555,581
2011	3,640,671
2012	3,747,344
2013	3,816,214
Total for 2007–2013	24,934,877

Source CDM project Ref. 1255—baseline and CER calculation

Table 5.3 Electricity generated, diesel displaced and CERs issued

Time period	Power generated (kWh)	Diesel displaced (liter)	CO_2e emissions avoided (ton)
May 13, 2008–August 31, 2009	4,243,315	2,705,137	3,395
September 1, 2009–May 31, 2012	8,461,518	6,182,441	6,768

Source CDM project Ref. 1255—monitoring reports 1 and 2

years of operations, delivering over 18.3 million kWh of electricity, displacing over 6.0 million liters of diesel and avoiding the emission of over 15,000 ton of CO_2 equivalent.[4]

No birds had been negatively affected by the wind energy project. It was observed that larger numbers of birds were hatching successfully (Neville 2008). The Ecuadorian post office launched a postage stamp to commemorate the successful implementation and operation of the project (Fig. 5.3).

> It should also be mentioned that the level of efforts put into the environmental assessment of this project and the way that the project is designed to return payments to the community through outreach, education and environmental stewardship is generally unique. For these reasons I feel the San Cristobal project is currently unique amongst the array of operating wind-diesel systems.[5]
> E. Ian Baring-Gould, Senior Mechanical Engineer, National Renewable Energy Laboratory, USA, 2 February 2009

The noteworthy aspects of the business model were as follows:

- All public and private sector project participants acknowledged the importance of the Galapagos Islands as an asset that belonged to all of humanity as a "world

---

[4] http://www.galapagoswind.org/, last accessed 9 December 2013.

[5] Schematic representation of the wind—diesel hybrid system on San Cristobal Island could be accessed at http://spectrum.ieee.org/energy/environment/wind-power-in-paradise.

Fig. 5.3 Postage stamp issued by the Ecuadorian post office to commemorate the inauguration of the project

nature area." "The e8 team approached this work with a level of caution akin to the curators responsible for da Vinci's Mona Lisa or Michelangelo's David."[6]

- The regulatory framework was modified to accommodate public funding of a project asset constructed and operated by a trust/corporation. Donations to the project by Ecuadorians were declared tax-deductible.

- Local involvement and support at community and governmental levels were extensive and continuous.

- Replacing generation capacity with cleaner sources and employing existing grid infrastructure, in itself, yielded very positive outcomes for the environment and for the economy.

- Given the remoteness of the project's location, the lack of relevant capacity among the islanders and the government's limited resources, wholly private initiative would have been impossible to implement.

- The project financing provided for competitive financial rates of return on the net-of-grant portion of the project costs. This model was being replicated on other islands within the Galapagos archipelago.

- Strong policy support in the form of permitting, environmental assessments, equipment import procedures, tax policies, tariff regulations, etc., has contributed immensely to the success of the project.

- Sustained involvement of the local implementation partner ("constructor") was important for overcoming the logistical challenges faced.

---

[6] News tab under http://www.eolicsa.com.ec/index.php?id=23, last accessed 9 December 2013.

Teaching Note

Case Synopsis

A large proportion of electricity on the island of San Cristobal was generated employing diesel. The island, forming part of the Galapagos Islands, was home to about 6,000 people and its fragile ecology and was especially vulnerable to oil spills during fuel transportation. Eighty-five percent of the island was part of the Galapagos National Park; the remaining 15 % was primarily agricultural land and the project was located on highlands outside the National Park area.

The San Cristobal Wind Power Project was a partnership among the Government of Ecuador, the United Nations Foundation and a specially constituted Wind Power Commercial Trust instituted by the e7 (later e8). e7, (www.e7.org, later renamed e8) was a coalition of nine electricity supply companies from the G7 countries (later G8) (see Footnote 1). The e7 Fund for Sustainable Energy Development was a nonprofit institution recognized as a non-governmental organization (NGO) with special consultative status, by the United Nations Economic and Social Council.

The islands were considered a paradise for birds and other amphibian and aquatic creatures and the international community was firmly committed to conserving the pristine ecosystem. The wind energy generation project was to go a long way toward this end. Among other things, the revenues from the sale of power generated were reinvested toward conservation measures, toward cleaner transportation as with battery-operated vehicles, and toward generating sustainable livelihoods for the local population.

Case Focus

The case is intended to highlight the working of partnerships involving diverse agencies: electricity utilities, multi-lateral development organizations, sovereign governments, and local communities.

Teaching Objectives

- Enabling students appreciate the (ironic) environmental challenges to be overcome to successfully implement a cleaner energy project.
- Identify the incentive structures faced by the participants of the grand coalition and success achieved with their alignment toward achieving project objectives.
- Highlight the importance of recognizing and respecting the rights of other life forms on the planet and to move the course participants away from an anthropocentric view of the world.

Case Objectives and Use

The case highlights the role of partnerships in achieving project goals. Even as the utilities involved offered technical and managerial assistance on a *pro bono* basis, the lead implementing agency viewed the project as an opportunity to enter a new market. Implementing a relatively small capacity project in a remote location did not receive an enthusiastic response from most mainstream wind turbine equipment manufacturers. Yet, MADE of Spain decided to play along and had also agreed to monitor the project remotely. The equipment vendor was also contracted to help build operation and maintenance (O&M) capacity among local technicians. The trust constituted to build and own the project was to redirect revenues toward conservation of native species of birds, especially the Galapagos Petrel. The Ecuadorian government benefited from avoiding import of diesel and hence from conserving foreign currency. Additionally, the project was slated to paint a greener image of the island(s) and to attract researchers and tourists in larger numbers. Thus, each entity had an interest in the project and each was working toward facilitating the other in successful implementation.

Teaching Plan

The case presents an interesting conundrum relating to the challenges involved in meeting environmental goals and economic targets with the use of environmentally sensitive technology, and yet having to invest long years into due diligence on the very environmental impact of the project.

The instructor could guide course participants to analyze the project from a non-anthropocentric perspective that man was merely one of the habitants of the planet and had to share space and rights with other living creatures. This could then be coupled with an economic perspective of the project, where the reduction in fuel bills, on electricity generation costs and the influx of tourists would by themselves justify implementation. Course participants could discuss the merits of incurring additional capital expenditure on the composite project involving among other things, expanding the jetty to unload the equipment, building long and winding roads around nesting sites, the transmission lines and interface with existing generation capacity, etc.

What Happened Next

The Ecuadorian government hoped to reach zero fossil fuel use by the year 2015. Some of the projects proposed for implementation on other islands could have included energy storage infrastructure: mini-hydro-pumped storage, geothermal systems, etc.

In an unrelated development, although involving wind-energy generation, in November 2013, a settlement was announced involving two wind farms in the USA state of Wyoming. Duke energy, the wind farm operator was to pay USD 1.0 million in settlement against the killing of 14 golden eagles (besides several birds of other species) over the preceding three-year period. The company had pleaded guilty under the Migratory Bird Treaty Act. This was the first such case against a wind-energy company. The fined amounts were to be paid to conservation groups including the North American Wetlands Conservation Fund, the Wyoming Game and Fish Department, and the National Fish and Wildlife Foundation. The company also promised to take appropriate measures to prevent bird deaths in the future (Fitzsimmons 2013). This development helps revisit the pricing of the San Cristobal project, places a value on native ecosystems, and helps justify the time and effort invested in environmental due diligence for the project.

References

Fitzsimmons EG (2013) Wind energy company to pay $1 million in bird deaths, The New York Times, 22 Nov 2013. http://www.nytimes.com/2013/11/23/us/wind-energy-company-to-pay-1-million-in-bird-deaths.html?_r=0. Accessed 2 Jan 2014

Neville A (2008) Top plants: San Cristobal Wind Project, Galapagos Islands, Ecuador. Power Mag 152(12):5

Chapter 6
Recogen, Sri Lanka: Thinking Out of the Shell

Risk Assessment

> ... *Haycarb and Recogen have shown us what local engineering skills and ingenuity can accomplish on their own. It's surely something that Sri Lanka can crow about, something which should make us want to stand up and cheer!*
> —Anne Abayasekara, The Island Online, 15 March 2009

Introduction

The Recogen, coconut shell to charcoal project in Badalgama, in Sri Lanka's northwestern province, involved the construction of a mechanized charcoaling plant for converting abundantly available coconut shell to charcoal using gasification cum pyrolysis—the thermo-chemical decomposition of organic material at high temperatures in an oxygen-deprived environment—and subsequently using the energy of the released gases and steam to generate electricity (Fig. 6.1).

The coconut shell charcoal was then crushed into a fine powder and mixed with a binder to form briquettes of the desired sizes and shapes. The generated electricity was to be exported to the Ceylon Electricity Board's grid under a power purchase agreement (PPA).

Project Description

The process commenced with crushing the raw coconut shells. The raw shell was then fed to a dryer that used waste heat of the flue gases to reduce the moisture content. The dry shells were fed to the processing kilns which would convert the shell to charcoal at temperatures of about 500 °C, over a 2-h period. Further downstream, the gases generated in the kilns were combusted and the heat used to generate steam which turned steam turbines, thereby generating electricity. The wet steam from the turbine was condensed and the recovered water was recycled to the boiler.

© Springer India 2015
S. Sunderasan, *Cleaner-Energy Investments*,
DOI 10.1007/978-81-322-2062-6_6

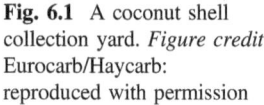
Fig. 6.1 A coconut shell collection yard. *Figure credit* Eurocarb/Haycarb: reproduced with permission

Prior to the implementation of the project, coconut shells were burnt in open pits to convert them to charcoal. It was estimated that a total of 185,000 t of shell was converted to 55,000 t of charcoal each year, representing a conversion efficiency of 29.73 %. More importantly, traditional charcoaling leads to the release of injurious flue gases into the atmosphere, leaving adverse impacts on the immediate neighborhoods and potentially contributing to the greenhouse effect at the global level. The proposed project, at full capacity, was scheduled to convert 77,442 t of coconut shells to 26,400 t of charcoal (34 % conversion efficiency), while trapping the flue gases and generating 5.8 MW of electricity and contributing to the mitigation of climate change as well. The business model is illustrated in the flowchart forming Fig. 6.2.

As of the year 2013, Sri Lanka was one of the largest producers of coconut in the world. Charcoal from coconut shells could be used for domestic or commercial cooking, barbecueing, or any other purpose as a replacement for naturally occurring and mined coal (Coconut Development Authority 2013). The country exported 6,919 metric ton of coco-char during calendar 2012 to some 36 countries, led by The Netherlands and Denmark, followed by Japan, Vietnam, and Germany, totally worth USD 3.70 million at an average traded price of USD 1,871 per metric ton.[1]

The Recogen plant received the "Development of Eco-materials/Eco-friendly Processes for Industry" award at the Sri Lankan National Science and Technology Awards in December 2007. It was then referred to as the world's only pollution-free charcoaling plant that captured and converted fugitive gases and vapors into electricity. The high-grade charcoal was supplied to parent company Haycarb's activated carbon manufacturing operations (Haycarb 2013a, b).

Recogen was a 100 % subsidiary of Haycarb whose equity investment stood at Sri Lankan rupee 370 million (~ USD 3.7 million). As of the fiscal year ending March

---

[1] http://ucap.org.ph/news-and-events/sri-lanka-coconut-shell-charcoal-export-up-in-2012, last accessed 5 November 2013.

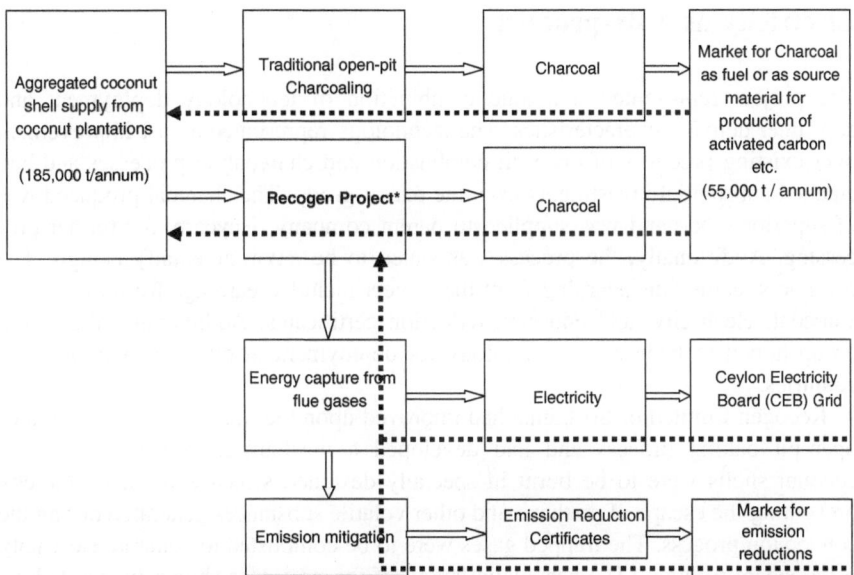

Fig. 6.2 The Recogen coconut shell to charcoal project depicting the material flow (⟹) and cash flow (■ ■ ■ ■ ▶): prepared by the author. * 8 kilns/77,442 t per annum of coconut shell/26,400 t of charcoal/5.8 MW of electricity at full capacity (Phased expansion from 1 kiln to 8 kilns and 1 MW to 5.8 MW over a 5-year period)

2011, Haycarb had recognized an impairment of the value of its investment, and Recogen's worth was stated at Sri Lankan rupee 220 million (~USD 2.2 million) based on a post-money valuation involving plant operation for 260 days a year, and cash flows discounted at 14 %.[2] Haycarb (www.haycarb.com) was a part of the Sri Lankan conglomerate Hayleys (www.hayleys.com) and had been a pioneer in the manufacture of activated carbon. The company had spent 10 years in the development of the new, now patented, process and was now able to export the technology to other coconut-growing countries. In addition to the commercial aspirations, the company also had a social perspective to its operations. Haycarb donated a large quantity of activated carbon powder to the Sri Lankan Ministry of Health to be given to patients as an antidote to people swallowing certain seeds while attempting to commit suicide! Activated carbon was also used in water filters, preservation of fruits and vegetables, kidney machines, odor absorbers, dechlorination in breweries and packaged water factories, municipal water treatment, aquariums, desalination plants, electroplating, recovery of gold from ore, air-conditioning, air purification in aeroplanes and submarines, and much more (Abayasekara 2009).

---

[2] Haycarb Annual Report 2011/12, p. 75 of 104.

Electricity as a By-product

The project represented a unique combination of technology deployment and consumer demand characteristics. The technology represented a vast improvement over existing practices of open-pit combustion and charcoaling processes and had entailed substantial investments from the parent group. The charcoal produced was of superior grade and was supplied to parent company, Haycarb, for further processing. Additionally, the process was slated to be environmentally benign. The revenue streams thus accruing from the project included earnings from the sale of charcoal, electricity, and emission reduction certificates. Additionally, the parent group hoped to license the technology for deployment in other coconut-growing countries.

Recogen Limited of Sri Lanka had improved upon the traditional small, outdoor open-pit baking process and had developed a mechanized process where the coconut shells were to be burnt in specially designed sealed equipment, thereby preventing the escape of methane and other volatile substances generated during the conversion process. The trapped gases were to be combusted to generate electricity using steam turbines. The schematic layout of the project is shown in Fig. 6.3.

The Recogen coconut shell charcoaling project was registered as a CDM project (Ref. 2364) in March 2009. The project had received 100 million Sri Lankan rupee ($\sim$USD 1.0 million) through the Environmentally Friendly Solutions Fund (E-Friends), which in turn was supported by the JICA (2013). In December 2011,

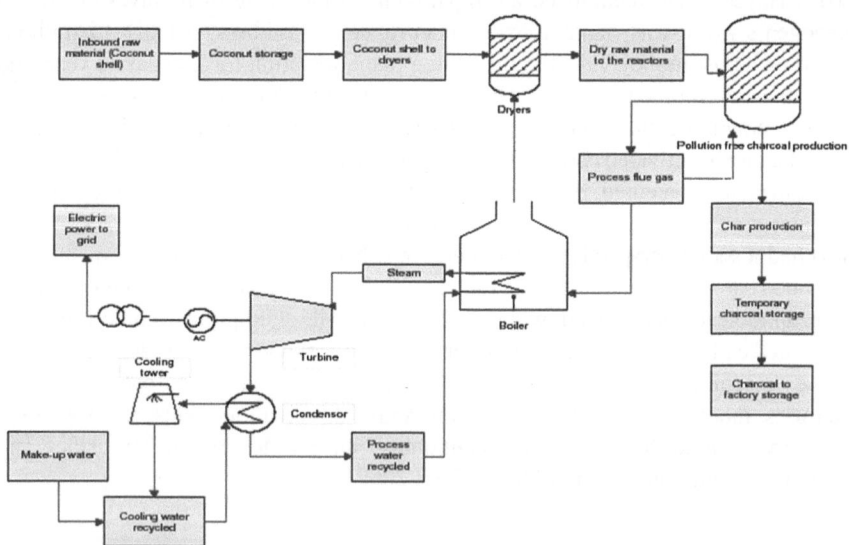

Fig. 6.3 Schematic lay out of the coconut-charcoal production and power generation process. *Data Source Haycarb* CDM Project Ref. 1463—Project Design Document, p. 8; Reproduced with permission

Table 6.1 Performance of Recogen, coconut shell to charcoal project in Badalgama, Sri Lanka

Monitoring period	Power generation (MWh)	Raw coconut shell consumed (tonne)	Charcoal produced (tonne)	Emission reduction achieved (tCO$_2$e)
April 2009–March 2010	2,629.74	12,598.365	4,247.884	4,747.00
April 2010–December 2011	5,887.36	33,030.634	9,507.828	10,341.00
January 2012–December 2012	3,760.68	22,786.410	6,804.930	7,232.00

Data Source CDM Project Ref. 2364—monitoring reports

the company traded-in 4,000 CERs it had received for the pollution-free charcoal production and power generation during 2009–2010 (Haycarb 2013a, b). The total power generated during calendar 2012 stood at 3,760.68 MWh compared to 5,887.36 MWh during April 2010–December 2011, and 2,629.74 MWh in April 2009–March 2010 (Table 6.1). An average of 5.6 kg of coconut shell was consumed per kWh of electric power generated. In its published annual report for the fiscal year 2012–2013, Haycarb had referred to Recogen a viable economic model that had excelled in manufacturing green charcoal.[3]

Very significantly, Recogen made the following statement with reference to displacing the competition: "In order to counter the loss of income of current charcoal manufacturers, due to project activity, Recogen would purchase the coconut shells from same pit charcoal suppliers at reasonable prices. Therefore, the project activity would not create additional demand for coconut shells but substitute the open-pit charcoal manufacturing process." Apparently, the suppliers of coconut shells are distinct from those who supply pit charcoal and the company had inadvertently placed itself directly in competition with the latter group.[4] According to the monitoring report submitted for the period January–December 2012, the commissioning of 5 additional kilns of 3,300 t/year and expanding the power generation capacity from 1.25 to 5.0 MW were deferred owing to limited access to coconut shell, as sizable quantities were still being procured by the open-pit operators.[5]

The noteworthy aspects of the business model are as follows:

- The project was a win–win–win proposition, using a locally available residual raw material to generate power for a power-starved nation, and supplying a crucial raw material for further processing and export and simultaneously contributing to social and environmental causes.
- The project was not able to expand beyond 9,900 t per year (3 kilns of 3,300 t per annum commissioned as of January 2008) of charcoal production and

---

[3] Haycarb PLC, Annual Report for the year 2012/2013, p. 12.

[4] CDM Project Ref. 1463—Project Design Document, p. 3.

[5] CDM Project Ref. 1463—Project Monitoring Report, 23 February 2013, pp. 3–4 of 27.

1.25 MW of power generation capacity, as the company found it difficult to displace the traditional open-pit operators and secure larger quantities of coconut shells. Consequently, the proposed 5-MW turbine and 25-t-per-h boiler were delayed to between 2014 and 2016.

- Supplanting the process and supplanting the people are different things. The company has sought to supplant existing operators rather than collaborating with them and attempting to upgrade their processes. It cannot be said with any degree of certainty if such an approach was attempted.
- The parent company's primary consideration was the availability of raw material to manufacture activated carbon, while power generation and GHG mitigation were secondary objectives. Yet the revenues contributed to the economic sustainability of the Recogen operation.
- The parent company had invested ten years and about a billion Sri Lankan rupees toward developing the process, which was now commercially exploitable and export-ready for deployment in other developing/coconut-growing countries.
- The electricity generated was fed to the grid and hence was likely to suffer the technical and commercial losses otherwise faced by the utility grid.

Employing personnel from the open-pit charcoal processing units that were forced out of business by the opening of the Recogen plant could have been a social positive for the project.

Teaching Note

Case Synopsis

The Recogen coconut shell to charcoal project in Badalgama, northwestern province, Sri Lanka, generated electricity as a by-product of the charcoal production process. The parent company, Haycarb, had spent close to a decade in developing the technology and was now ready to export it to other coconut-growing countries. The mechanized charcoaling plant converted coconut shell to charcoal through gasification and thermo-chemical decomposition of the organic matter in an oxygen-deprived environment. The electricity was generated by recovering the waste heat from the flue gases and then running a steam turbine. The steam was condensed and the water was recycled with minimal round-trip losses. The high-grade charcoal was supplied to the parent company for the production of activated carbon which found application across several sectors.

The key-distinguishing features of the business model were the fact that the main product was charcoal, while electricity itself was a by-product. The electricity so generated was fed to the national grid operated by the Ceylon Electricity Board. Between April 2009 and December 2012, the project had generated over 12,276 MWh of electricity. The charcoal had a ready captive market within the business group.

The mechanized process was to supplant traditional open-pit charcoaling which allowed the fugitive gases escape into the atmosphere, contributing to the greenhouse effect. While the project was a vast technological improvement over existing processes, the promoters seem to have underestimated the existing social and business relationships between the coconut shell wholesalers and the open-pit operators.

Case Question

The case is intended to highlight the importance of existing social structures, the roles of the agencies involved, and the positioning of emergent technologies, within an existing scheme of things.

Teaching Objectives

- Enabling students appreciate the social aspects of cleaner energy projects.
- Identify the lessons learnt and the corrective action necessary.
- Revisit the project assumptions and remodel the project to try and blend with existing social structures.

- Make subjective evaluations of risk and make recommendations on the viability of the proposed ventures and the qualifications or reservations associated with such viability.

Case Objectives and Use

The case focuses on identifying sources of project risk and on helping redesign certain aspects of the project, and on appropriately managing such risk, *inter alia,* entering into partnerships and helping existing operators scale-up with the new technology. In this case, however, the new technology was developed in house and the business model involved bypassing existing operators altogether. The project was to source coconut shell from the wholesalers, possibly putting existing open-pit operators out of business. However, the linkages between the wholesalers and the existing entrepreneurs were too strong for the company to disrupt and to scale up as proposed.

The company was compelled to delay the commissioning of five additional kilns of 3,300-tonne-per-year capacity and hence the expansion of power generation capacity from 1.25 to 5.0 MW, because of constrained raw material supply. Consequently, the proposed installation of the 5-MW turbine and the 25-tonne-per-h boiler were put off to between 2014 and 2016.

Teaching Plan

The case presents a project launch scenario where a leading business group had developed a new technology and sought to create a supply chain to serve its ongoing activated carbon business. In the process, however, the company seemed to have misconstrued the roles of the wholesalers who dealt in coconut shell—the primary input material—and fellow charcoal makers.

Technology development is intrinsically risky and could easily suffer time and cost overruns. The parent company had invested a billion Sri Lankan rupees and about a decade's worth of time toward developing this technology. Yet this put the company directly in competition with relatively low-technology rivals engaged in traditional practices. It is not clear whether the company hoped that adequate raw material would be available to supply all players or whether some of the traditional open-pit operators would exit the industry.

The instructor could help evolve scenarios and induce role play among the course participants to analyze the company's *ex ante* strategy formulation in this context, commencing with a threadbare analysis of the statements below:

... in order to counter the loss of income of current charcoal manufacturers, due to project activity, Recogen would purchase the coconut shells from same pit charcoal suppliers at reasonable prices. Therefore the project activity would not create additional demand for coconut shells but substitute the open-pit charcoal manufacturing process.

Read with, "Impact on Society: Haycarb interacts directly with over 200 local charcoal suppliers. Through its enterprise, Haycarb provides a livelihood and enriches lives far beyond the direct supplier, reaching many lives from the plantation growers, the farm hands, to those engaging in charcoaling and beyond. We are thus an indirect employer touching thousands of lives each day."[6]

1. Which agencies would constitute the project's competitors? Complementors? Do the course participants observe that the project had underestimated the strength of the competition? Complementors?
2. Rework the financial model to factor in a delay of 5 years in the scale-up.
3. If the project were any less attractive owing to the delay in scale-up, what strategies would course participants recommend to make the project more viable?

What Happened Next

The Recogen project produced a quarter of all of Haycarb's charcoal needs. The company believed that backward integration to the production of charcoal had given it greater control over the quality of its main input material. In keeping with the company's primary focus on generating feedstock for the activated carbon business—with electric power remaining a by-product—the company believed that it was better positioned to customize its activated carbon offering to meet customer specifications by tuning the charcoaling process itself, as opposed to sourcing charcoal from the market and then working it to suit customer requirements (Feature: Eurocarb Products Limited 2010). The process was slated to yield charcoal of consistent quality and reliable yield even in damp and adverse weather conditions.[7]

References

Abayasekara A (2009) Three cheers for a uniquely Sri Lankan Project—Recogen Pvt. Limited, 15 March 2009. http://www.island.lk/2009/03/15/business5.html. Accessed 5 Nov 2013
Coconut Development Authority (2013) Ministry of Coconut Development and Janatha Estate Development, Government of Sri Lanka, www.cda.lk. Accessed 4 Nov 2013

---

[6] Haycarb PLC Annual Report 2011/12, p. 51 of 104.

[7] "Green Carbon", http://www.eurocarb.com/green-carbon/, last accessed 2 January 2014.

Feature: Eurocarb Products Limited (2010) Filter media: new process points way forward for activated carbon. Filtration + Separation, pp 39–41 (March/April 2010)

Haycarb (2013a) Haycarb trades carbon credits under Kyoto protocol, *Financial Times*, 8 Dec 2011. http://www.ft.lk/2011/12/08/haycarb-trades-carbon-credits-under-kyoto-protocol/. Accessed 5 Nov 2013

Haycarb (2013b) Haycarb's Recogen plant wins national science and technology award, The Sunday Times Online, 23 Dec 2007. http://www.sundaytimes.lk/071223/FinancialTimes/ft338. html. Accessed 5 Nov 2013

Japan International Cooperation Agency (2013) First ODA backed private sector project registered as CDM Project, Press Release 31 Mar 2009. http://www.jica.go.jp/english/news/press/2009/090420.html. Accessed 5 Nov 2013

Chapter 7
Delhi Airport Metro Line: Aborted Takeoff?

Risk Assessment and Management

> *Unlike the government, private firms have to depend on high cost finance*
>
> —Mamuni Das
> The Hindu Business Line, 29 July 2013.
>
> *All parties involved in the project should have done more due diligence at all levels*
>
> —Sudhir Krishna
> Chairman Delhi Metro Rail Corporation and Secretary, Ministry of Urban Development, Government of India.
> The Hindu Business Line, 29 July 2013.

Background

The 22.70-km Delhi Airport Metro Line costing close to USD 1 billion (INR 5,700 crore; 1 crore = 10 million) was to ferry 42,000 passengers each day between Delhi's upgraded and enlarged airport, the proposed "Aerocity," and the New Delhi Railway Station. As of mid-2013, actual average daily passenger numbers stood at about 13,500 causing the operator, Delhi Airport Metro Express Private Limited (DAMEPL), an estimated loss of about USD 700,000 (INR 4 crore) a month (Das 2013).

In a public notice displayed on its website, DAMEPL had claimed that the company had terminated the concession agreement to operate the line in October 2012 owing to "substantial defects in the civil structure"[1] and had served a final notice of cessation of Metro Line operations as of the July 1, 2013. "Since Reliance Infra [DAMEPL's parent] was unable to run the line, DMRC [Delhi Metro Rail Corporation and contracting agency] started operations in the larger public interest,"

---

[1] http://www.delhiairportexpress.com/home/index.html, last accessed August 19, 2013.

© Springer India 2015
S. Sunderasan, *Cleaner-Energy Investments*,
DOI 10.1007/978-81-322-2062-6_7

in the words of Mr. Sudhir Krishna, Secretary, Urban Development and Chairman of the DMRC Board of Directors (Roychowdhury and Chakravartty 2013). DAMEPL had made a claim for 130 % of its equity contribution and 100 % of its debt, collectively amounting to about USD 490 m (INR 2,800 crore).

The Project Company

Delhi Airport Metro Express Private Limited (DAMEPL: www.delhiairportexpress. com) was the special purpose vehicle (SPV) designated to operate the Delhi Airport Metro Express Line connecting Delhi city's Indira Gandhi International Airport with the city center at the New Delhi Railway Station. The SPV was jointly owned by Reliance Infrastructure Limited of India and CAF of Spain. Reliance Infra, part of the Reliance Group, has been engaged in real estate projects, roads and bridges, and the generation, transmission, and distribution of power. Reliance Infra was also involved in the implementation of two metro lines in Mumbai (formerly Bombay). *Construcciones y Auxiliar de Ferrocarriles, S.A* (CAF) of Spain specialized in the design, manufacture, supply, and maintenance of equipment and railway system components, with production facilities in Spain, in the USA and in France (Fig. 7.1).

Project Description

The 22.70-km Delhi Airport Metro Express was built with a view to providing "safe, swift, reliable, and eco-friendly" transport between central Delhi and Delhi's Indira Gandhi International (IGI) Airport. The services were benchmarked against

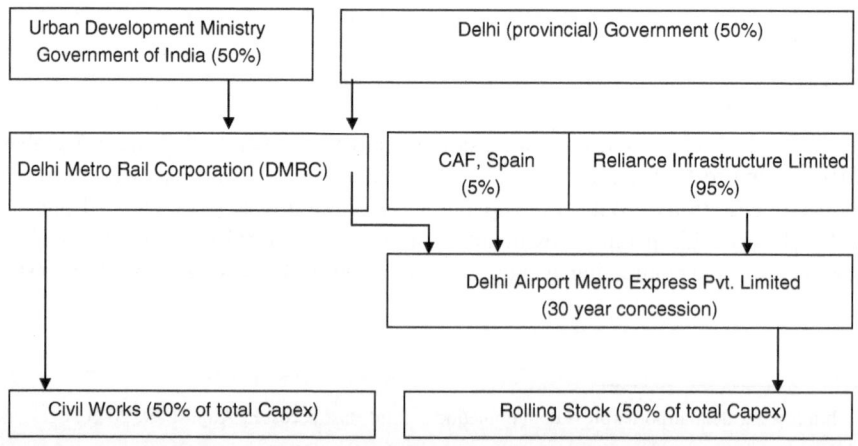

Fig. 7.1 Ownership structure and funding pattern for the Delhi Airport Metro Project

Table 7.1 Technology providers for the Delhi Airport Metro Express

Signaling, power transmission, baggage handling system	Siemens mobility
Communications systems	Alcatel
Ticket machines	Indra systems
Platform screen doors	Faiveley
Control and automation for station management system	Bluestar/Honeywell
Six-car trains (8 numbers)	CAF Spain

http://en.wikipedia.org/wiki/Delhi_Airport_Metro_Express, last accessed August 20, 2013

the best in the world and the line aspired to rank among the top five airport link services in the world.[2] The vendors chosen for the supply of equipment are listed in Table 7.1. The dedicated—elevated and underground—corridor was intended to decongest roads and shorten travel times substantially. Three of the six stations *en route* were to offer "city airport terminal" facilities including baggage check-in for passengers.

The metro line was slated to displace a large number of personal transport vehicles plying on the airport–city route and thereby contribute to reducing air and noise pollution in the city. The trains were scheduled to run at 15-min intervals for 18 h each day between 530 and 2,330 h. The airport line was integrated to provide for seamless interchange with the existing and projected DMRC metro network, providing access links to large parts of Delhi city.

A single journey between New Delhi Railway Station and the IGI Airport was priced at INR 150 (~USD 2.6). The airport line hoped to provide a "complete shopping and dining experience to the people visiting the stations." The Airport Express was creating premium retail spaces (67 outlets) at two of its stations located in the city center. The retail spaces were to be accessible to non-commuters as well as to commuters. The project was also expecting to generate revenues from the sale of advertising and product display space on station facades and inside the stations, on the luggage trolleys, on the smart cards, and on the trains themselves, including the branding of entire trains. In addition, paid porter services were offered to help passengers with their luggage and a tie-up with a cab service provided links to various parts of the National Capital Region within and beyond Delhi.

Project Cost and Means of Finance

The total up-front cost of the project was estimated at INR 5,700 crore (~USD 1 billion at 1USD = INR 57). As the entire line was built with ballast-less track, with the rails resting on rubber pads and concrete sleepers, the permanent way alone had cost INR 247.82 crore per kilometer, 40 % higher compared to other Delhi Metro Lines that cost INR 175 crore per kilometer (Chakravartty 2012). Further, the

---

[2] Vision and Mission of the Delhi Airport Metro Express.

wheels came equipped with a flange lubrication system to reduce noise and to enhance riding comfort.

The government through the DMRC was to contribute 54 % of the capital expenditure, while the SPV was to bring in the residual 46 %. DAMEPL, the SPV, was funded with 30 % in equity and 70 % in debt. It was reported that a consortium of 12 banks led by Axis Bank Limited had provided debt financing of INR 2,000 crore (~USD 350 m). The company had created total assets worth INR 2,244.62 crore and would need an additional INR 946 crore invested into operations and maintenance over the 30-year concession period.

Project Strengths and Risks

The airport line was to ferry 40,000 passengers a day, with the number escalated at 7.3 % per annum, reaching 86,000 persons per day in the year 2021. The numbers were based on the assumed construction of the Aerocity: city-side development associated with the expansion and modernization of the IGI Airport.

Construction deadlines set to facilitate movement of passengers during and around the October 2010 Commonwealth Games were missed, and the line commenced commercial operations by February 2011. However, by January 2013, average daily passenger numbers stood at 10,000, mandating an increase in fares by INR 30, further worsening the attractiveness of the metro line to passengers (Roychowdhury and Chakravartty 2013). In the 18 months to end of the September 2012, the line had carried 6.8 million passengers. Apparently, the passenger traffic which was to contribute 25 % of project revenues was overestimated. Additionally, commuters found the line inaccessible and potentially unsafe at nights. The airport metro line connected to terminal 3 at the IGI Airport, and accessing other terminals was not easy (Chakravartty 2012). Revised estimates for projected passenger numbers for the year 2021 were much lower at 35,741 persons a day.

The airport line had reported defects in 540 (over 90 % of the total) bearings that supported the track. The Commission for Railway Safety had directed the DMRC— the agency responsible for the construction of the civil structures, viaduct, tunnels, and stations—to conduct load tests on several girders under different conditions. Passenger services were suspended between July 7, 2012, and January 22, 2013. After the repairs were taken up, the speed of the metro trains was revised from 50 km per hour (40-min travel time) to 80 km per hour, as opposed to the originally promised 120 km per hour/18-min travel time.

Pending a resolution of the legal issues, the DMRC took up running the airport line to ensure uninterrupted services for the commuters. Monthly operating expenses were estimated at INR 7.0 crore against revenues of INR 3.0 crore yielding a net loss of INR 4.0 crore a month. After DAMEPL's withdrawal from the project, DMRC had proposed to augment secondary sources of revenue from city-side developments at the metro stations.

Ridership had stagnated at around the 10,000 persons a day mark (DMRC 2013). By September 2013, the Central Industrial Security Force (CISF) responsible for providing security on the airport express line expected to recover its dues of about INR 16 crore (INR 160 m $\sim$ USD 2.8 m) from disposing off security equipment including baggage scanners, metal detectors, etc., supplied by the project company (Haidar 2013). The employment prospects of some 500 people employed by DAMEPL were yet undecided.

Notwithstanding the specific agency charged with operating the airport line, fundamental issues relating to viability of the business model and challenges associated with lower than projected ridership remained (Delhi Metro 2013).

Data assembled from parent company annual reports

Balance sheet					Re. crore
	31-Mar-09	31-Mar-10	31-Mar-11	31-Mar-12	31-Mar-13
Share capital	373.91	466.96	0.01	0.01	
Reserves and surplus			−15.47	−341.13	
Debts + current liabilities + deferred tax liability	10.52	570.37	2,482.39	2,844.32	
Total liabilities	384.43	1,037.33	2,466.93	2,503.2	
Fixed assets + capital work in progress + current assets + deferred tax asset	384.43	1,031.32	2,244.62	2,502.65	
Mutual fund investment	0	6.01	222.31	0.55	
Total assets	384.43	1,037.33	2,466.93	2,503.2	
Profit and loss account					
Revenues (including other income)			7.01	38.89	
Cost of sales			−22.48	−364.54	
Earnings before tax			−15.47	−325.65	−92.05
Provision for taxation			0	0	0
Earnings after tax			−15.47	−325.65	−92.05

Parameter	Unit	Value	Remarks
Return on equity expected	%	12.75 %	Reliance Mumbai Metro One/CDM Project Document: http://cdm.unfccc.int/Projects/DB/SQS1302526267.27/view
Concession period	Years	30	
Cost of debt (domestic)	%	11.25	Reliance Mumbai Metro One/CDM Project Document: http://cdm.unfccc.int/Projects/DB/SQS1302526267.27/view
Cost of debt (foreign currency)	%	7.50	
Tax rate	%	33.99	30 % tax + 10 % surcharge + 3 % cess

Teaching Note

Case Synopsis

The high-speed rail link between New Delhi's upgraded and enlarged Indira Gandhi International (IGI) Airport and the city center was forecast to ferry 40,000 passengers each day, on average. The proposed "Aerocity," the city-side development at the airport, was vitally important to drive passenger volumes. Retail businesses at stations and advertisements on trains, platforms, etc., were to garner significant portions of projected revenues. Retail and promotion-related revenues, in turn, depended on passenger numbers which averaged about 25–30 % of original estimates.

The permanent way of the line and the civil structures was built by the Delhi Metro Rail Corporation (DMRC), while the Reliance Group's Reliance Infra partnered with CAF of Spain to provide the rolling stock and to operate and maintain the commuter service. It was estimated that the service was earning about INR 3 crore (INR 30 m ∼ USD 0.5 m) in revenues each month, while expenses stood at about INR 7 crore (INR 70 m ∼ USD 1.18 m), thereby yielding a net loss of INR 4 crore (INR 40 million) each month, not counting interest payments on project debt.

The private operator had pointed to defects in the civil structure, and joint surveys did indeed reveal shortcomings. By July 2013, the operator abandoned the line and DMRC-restored operations in the interest of the commuters. DMRC believed that the termination was more of an economic decision rather than a technical one. The intransigence seemed to reflect insufficient due diligence investigations by all the agencies involved.

Case Question

The case is intended to highlight the importance of assumptions and the validation of the assumptions, predominantly those relating to project costs and revenue models. The case also deals with the significance of establishing the roles and responsibilities of the public and private partners involved in joint sector projects.

Teaching Objectives

- Enabling students appreciate the merits of detailed due diligence investigations.
- Identify the lessons learnt and the corrective actions to be taken.
- Revisit the project assumptions and remodel the projections with more conservative estimates of ridership, revenues, etc.

- Taking cognizance of various quantitative inputs and subjective evaluations of risk and returns and making recommendations on the viability of the proposed ventures, and the qualifications or reservations associated with such viability.

Case Objectives and Use

The primary objective of the case is assessing project risk and designing a risk management plan, *inter alia*, allocating risks to the parties best able to manage them. In the normal course, the project hardware could be contracted out to a private partner on a build-own-operate model and the services could be provided by a public entity, possibly at subsidized prices. In this case, however, the civil structures were erected by the public entity and the service was offered by the private partner—hoping to recover economic costs in full—and who was ultimately required to bear the traffic and revenue risk. The traffic volumes were contingent on related developments, which had not materialized as planned.

A significant component of the sensitivity analysis is the circular relationship between passenger numbers and advertising revenue, even though the two streams were analyzed as stand-alone sources of project revenue. The willingness to pay for advertising on trains, platforms, stations, and even on tickets is driven by the anticipated lack of clutter that otherwise afflicts outdoor media, and the size and demographics of the target audience. When passenger numbers are far lower than originally anticipated, advertisers tend to scale back on their spends in tandem, thus magnifying the acute shortfalls in project revenues.

Teaching Plan

The case presents a cleaner transportation project, where the service provider needed to balance between financial returns earned on behalf of the shareholders and socioeconomic and developmental returns. Ironically, an infrastructure major was invited to operate a service, while a government body undertook construction of the civil structures and ancillary facilities. Normally, investors take on construction and commissioning risks, while tax payers, through the public body, essentially absorb the risks of uncertain and delayed cash flows generated by the infrastructure project. Among other things, the capital structure and the cost of debt determine the solvency of the project in the face of sharp declines in traffic realizations.

The instructor could guide course participants to analyze the financial projections for the project and to draw up scenarios, where a private entity would be incentivized to operate the service, as originally intended.

The instructor could guide the students to evaluate the project in a number of small steps as laid out hereinbelow:

1. Discuss the turbulent operating history of the Airport Express and list key dates.
2. Bring out the risk factors and the agencies best equipped to manage and hence mitigate them.
3. Draw parallels to stand-alone projects versus projects-forming parts of larger networks to benefit from network effects.

What Happened Next

The DMRC proposed to provide feeder bus services to each of the stations on the metro line to improve access; daily commuters were to receive huge discounts on the quoted rack rates; the frequency of services was also to be increased to help shore up ridership numbers (Pragya 2013).

For its part, Reliance Infra transferred 65 % of the shares in DAMEPL, the project company, to the Reliance Delhi Metro Trust, thereby converting the project company into an "associate" rather than a "subsidiary." Reliance Infra's share of the losses incurred by the project company stood at INR 311.13 (of 341.13) as of March 2012. Losses of 83.95 crore incurred by the parent during April 2012—March 2013, therefore, translated into business losses for the project company of INR 92.05 crore. The parent company had netted of the total losses of INR 395.08 crore against subordinate debts, as reported in its annual report for the year ended March 2013 (Reliance Infra 2013).

Studies were being conducted to ascertain whether the airport metro line could form part of a larger regional rail transit corridor (Dash 2013), Delhi–Gurgaon (Haryana)–Bhiwadi–Rewari–Alwar (Rajasthan) linking industrial areas and residential layouts and helping decongest New Delhi city.

References

Chakravartty A (2012) High-speed derailment. Down to Earth. http://www.downtoearth.org.in/content/high-speed-derailment. Accessed 20 Aug 2013
Das M (2013) Delhi airport metro line debacle: the way forward. Hindu Bus Line, p 5
Dash DK (2013) Airport metro express line may become part of Alwar link. The Times of India. http://articles.timesofindia.indiatimes.com/2013-09-04/india/41764664_1_airport-metro-express-line-delhi-metro-ud-ministry. Accessed 11 Sept 2013
Delhi Metro (2013) Delhi Metro Rail Corp Takes Charge of Airport Line www.rediff.com. http://www.rediff.com/money/report/delhi-metro-rail-corp-takes-charge-of-airport-express-line/20130702.htm. Accessed 11 Sept 2013
DMRC (2013) DMRC plans to monetize space on airport express line. The Financial Express http://www.financialexpress.com/news/dmrc-plans-to-monetise-space-on-airport-express-line/1157333

Haidar F (2013) CISF to sell equipment to recover 16 crore dues. Hindustan Times. http://www.hindustantimes.com/India-news/NewDelhi/CISF-to-sell-equipment-to-recover-16-crore-dues/Article1-1119403.aspx. Accessed 11 Sept 2013

Reliance Infra (2013) Reliance Infra Annual Report for the year April 2012—March 2013. http://www.rinfra.com/pdf/RINFRA_Full%20AR_2012-13_010813.pdf, p 52, 98, 137. Accessed 11 Sept 2013

Roychowdhury A, Chakravartty A (2013) Profitable exit. Down to Earth. http://www.downtoearth.org.in/content/profitable-exit. Accessed 10 Aug 2013

Singh P (2013) Measures a foot to increase airport metro ridership. The Pioneer. http://www.dailypioneer.com/city/measures-afoot-to-increase-airport-metro-ridership.html. Accessed 11 Sept 2013

Chapter 8
Godawari Green: Got-to-Worry?

The Time Cost Compact

...it is too early to brand [CSP] a success or a failure
Jasmeet Khurana, "The Verdict is Still Out," Sun & Wind
Energy, October 2012

...the CSP projects hang by a thread
Ankur Paliwal, "Sun Block," Down To Earth Magazine,
May 15, 2013

Background

In 2010, Godawari Power and Ispat Limited, through wholly owned subsidiary Godawari Green Energy Limited (ggelindia.com) ("Godawari," the Project Company), and six other firms were awarded contracts to install concentrating solar power (CSP) plants of the type shown in Fig. 8.1 and to sell electricity at an average price of INR 11,480 ($\sim$ USD 200) per MWh. The seven projects added up to 470 MW in capacity and involved about USD 1 billion in investments, averaging about USD 2.2 million per MW in capital costs.

Godawari's 50 MW CSP plant, claimed to be Asia's largest concentrating solar power plant implemented until then, located in the north-western Indian state of Rajasthan, comprised 5,760 mirrors that helped concentrate solar radiation and generated steam to drive turbines. The project asset was installed in Naukh village, district Jaisalmer, Rajasthan state. The generated power was to be sold to the NTPC Vidyut Vyapar Nigam (NVVN: nvvn.co.in), designated to procure power under the first phase of the Government of India's National Solar Mission.

The project involved the use of parabolic trough technology as shown in Fig. 8.2, most frequently used in other parts of the world, and with a proven record of robust performance.[1]

---

[1] http://www.hiragroupindia.com/companies/ggel.php, last accessed 14 August 2013.

© Springer India 2015
S. Sunderasan, *Cleaner-Energy Investments*,
DOI 10.1007/978-81-322-2062-6_8

Fig. 8.1 A view of a concentrating solar reflector module; *photograph* by the author

Fig. 8.2 A schematic diagram of a concentrating solar power array

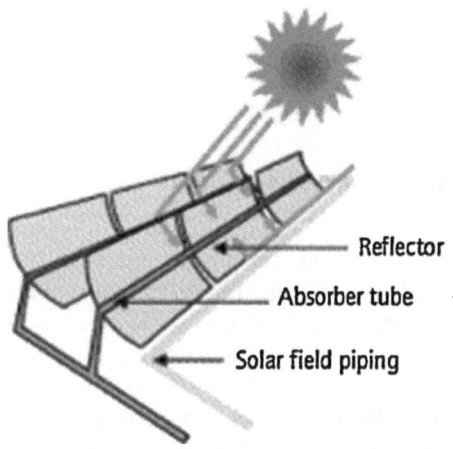

The plant was commissioned in June 2013, about one month behind schedule (Pearson 2013a, b). The contract mandated a tight implementation timeline which added to project costs. The harsh site conditions made matters worse, and consequently, all of the Rajasthan-based projects were running behind schedule. Given the higher capital expenditure involved and the fixed tariff for generated power, the plant's profitability was highly sensitive to the plant load factor (PLF) and hence to the quantum of power generated.

Analysts, investment bankers, and government officials observed that the bidders lacked reliable solar radiation data for the locations and had overlooked the possibility of dust deposits adversely affecting generation. The dust was also feared to be abrasive, reducing the efficiency of the mirrors and shortening replacement cycles. The Godawari project was further set back by a dust storm that caused a building at site to collapse.

Table 8.1 Project parameters, tariffs bid, and bank guarantees provided (Paliwal 2013; Singh 2013)

Size (MW)	Project proponent/special purpose vehicle	Parent company/group	Bank guarantee (Rs. crore)	Bid (Re/kWh)	Location (state)
100.0	Diwakar Solar	Lanco Infratech	183.80	10.49	Rajasthan
100.0	KVK Energy	Lanco Infratech	148.30	11.2	Rajasthan
50.0	Godawari Green Energy	Hira Group	52.70	12.2	Rajasthan
100.0	Rajsun Technies		114.70	11.97	Rajasthan
50.0	Corporate Ispat	Abhijeet Group	51.95	12.24	Rajasthan
50.0	Megha Engineering		71.40	11.31	Andhra Pradesh
20.0	Aurum Renewable Energy		21.20	12.19	Gujarat

Further, a weakening Indian rupee (INR 46.40/1 USD on the June 1, 2010 to INR 56.55/1 USD on the June 1, 2013) pushed up the costs of imported components by as much as 25 % relative to initial estimates. Some of the overseas suppliers exercised their dominant market position to raise prices for niche products such as the heat transfer fluid (HTF). Government policy on storage and transmission of power during the night or during off-peak hours was yet unclear.

None of the 470 MW of CSP projects listed in Table 8.1 met the May 2013 deadline. The commissioning of the next project was at least 6 months away. Some were delayed by over 15 months (Singh 2013). A few of the project companies had stalled construction, and work had not progressed beyond the foundation laying stage. It was beginning to appear that eventually, many of the proposed plants might never be built. Project promoters cited delays in the construction of the water pipeline by the provincial government of Rajasthan, delays in procurement of the HTF and other components sourced from overseas, and delays in securing financing (India Solar Weekly Market Update 2013). The delays could also be attributed to the lack of experience with the CSP technology in the country (Ahmed 2010).

Some of the claims made by the project promoters were indeed genuine. Access to financing was a key concern, as banks perceived solar energy investments to be highly risky, and there was no experience with CSP technology in the country. Water was scarce in the solar-radiation-rich, arid regions selected by project developers. Margins of error on available solar radiation data were high. The actual solar resource available was found to be 15–20 % lower than the Ministry's projections, and hence, plants had to be redesigned, procurement of components had to be delayed, and larger tracts of land had to be acquired. Power evacuation infrastructure was not readily accessible in such locations, and tight project deadlines made it difficult to switch sites.

Consequently, The Ministry of New and Renewable Energy (MNRE, "the Ministry"), Government of India, had recommended an extension of the implementation window by 10 months (Pearson 2013a, b). The penalty for delayed commissioning was originally set at INR 100,000/MW/day. The Ministry chose not to liquidate the guarantees provided by the developers for an additional 10 month window, i.e., until March 2014, foregoing some USD 42.50 million in penalties. Godawari had provided a guarantee of INR 52.7 crore (1 crore = 10 million) against performance defaults, relating to the commissioning and operation of the project. Table 8.1 also lists amounts provided in bank guarantees for the seven CSP projects.

The Project Company

Godawari Green Energy Limited, "the project company," was incorporated in 2009 as a subsidiary of Godawari Power and Ispat Limited (http://www.gpilindia.com/), "the parent company," with a view to generating ecologically sensitive electric power and to help bridge the demand–supply gap in the country. The companies formed part of the Raipur, Chhattisgarh state-based Hira Group (http://www.hiragroupindia.com/), founded in 1970, and engaged in various sectors including steel, ferro-alloys, sponge iron, cement, power, coal washing, mining and crushing, and pellets and electrical engineering products. Within the power sector, the group operated 8.1 MW of wind energy capacity and 146 MW of waste heat recovery systems and the foray into CSP was an effort to further diversify the business operations of the group.[2]

Over the ten-year period between March 2003 and March 2012, the parent company had grown revenues from INR 67.78 crore ($\sim$ USD 1.13 million at August 2013 rate of about INR 60/USD) and net profits of INR 1.50 crore ($\sim$ USD 0.25 million) to revenues of INR 1919.53 crore ($\sim$ USD 320 million) and a profit after tax of INR 78.95 crore ($\sim$ USD 13 million). The parent had consistently reported profits over the ten-year period 2002–2003 to 2011–2012.

The project company had no business income for the years ending March 2010, 2011, and 2012, and the income reported related exclusively to interest earned on fixed deposits held with banks. The balance sheets and income statements of the project company for the three years to March 2013 are annexed to this case document.

Project Description

CSP technology uses arrays of reflecting surfaces to concentrate heat from the sun into a small area. The heat is passed via a heat transfer fluid to a conventional power plant where it is turned into steam to drive a turbine. Parabolic trough collector

---

[2] http://www.hiragroupindia.com/products/index.php#power, last accessed 14 August 2013.

plants are made up of horizontal arrays of reflectors which track the sun. The HTF could attain temperatures as high as 4,000 °C. Most significantly, CSP plants can incorporate storage systems, most likely consisting of molten salts, enabling supply to continue after sundown (Hitchin 2013). "Using molten salt as the integrated working fluid allows for no loss in efficiency or heat during a heat transfer, and it does not require any natural gas for operations or steam consistency" (Muirhead 2013).

The 50 MW Godawari Green CSP project had executed a 25-year power purchase agreement (PPA) with the NTPC Vidyut Vyapar Nigam Limited, the government-owned and designated power purchasing agency (nvvn.co.in). The government's estimates of capital cost of project construction, which were employed in determining tariffs, and hence returns to the investors, were revised from INR 13 crore per MW ($\sim$USD 2.2 million) to about INR 15.3 crore per MW ($\sim$USD 2.55 million) by the Central Electricity Regulatory Commission (CERC) (Subramaniam 2010). The company's own estimated cost of the project was about INR 15.88 crore per MW ($\sim$USD 2.65 million). The PPA tariff was fixed at INR 12.2 ($\sim$US cent 20) per kWh.

Lauren Jyoti Private Limited (http://www.laurenjyoti.com/Project.aspx) was appointed the principal contractor for engineering, design, procurement, construction, manufacture, erection, commissioning, equipment testing, achieving COD and performance, and guarantee testing of the project. Vendors for major components are listed in Table 8.2. The schematic layout of the project is shown in Fig. 8.3.

Project Cost and Means of Finance

The capital cost of the project was estimated at INR 15.88 crore (INR 158.8 million, $\sim$USD 2.65 million) per MW, amounting to INR 794.0 crore or about USD 133.3 million for the entire plant. Operation and maintenance expenses for the

Table 8.2 Technology providers/vendors for major components; CDM—project design document (PDD)

Equipment	Vendor/technology provider
Steam turbo generator (STG)	Siemens, Sweden
Heat exchanger	Alborg CSP, Denmark
Solar collector loops	Design by SBP, Germany
Cooling tower	Paharpur Cooling Tower, India
Boiler feed pump	Sulzer India
Heat transfer fluid vessel	Ravi Industries, India
Deaerator	Ravi Industries, India
HTF pump	Sulzer India
Reflectors	Flagbag, Germany
Receiver tube	Schott Glass, Germany

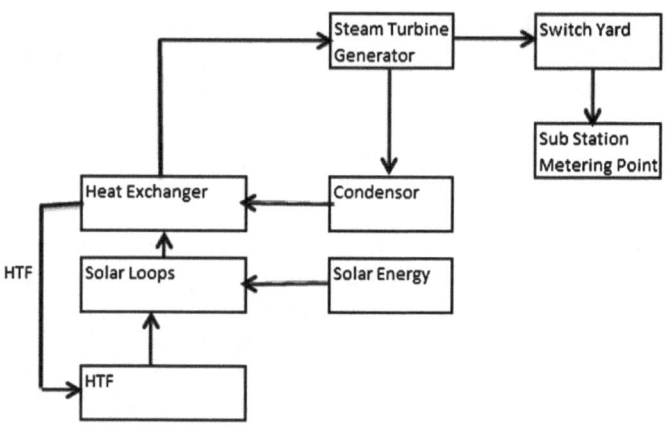

Fig. 8.3 Schematic diagram of the project; CDM—PDD, p. 7

first year were estimated at INR 137,400 per MW or about INR 6,870,000
(~USD 114,500) for the plant, further escalated by 5.72 % each year starting
March 2011. The project company was expecting to mobilize debt at 12.25 % with
a tenure of 12 years, and debt was to constitute 78.40 % of the project's capital
structure. The corporate income tax rate applicable, including base rate and education cess, was 33.22 %.[3] The return on equity would vary subject to the approval
of the project by the CDM executive board and with the unit pricing of emission
reduction certificates (CER: "carbon credits") (Purohit and Purohit 2010) and the
prevailing market prices for the CER.

Project Strengths and Risks

India was ranked eighth among 40 countries on the Ernst & Young Renewable
Energy Country Attractiveness Index, on the prospects for growth in the onshore
wind energy sector (Ross 2013). India, however, managed to retain its fifth position
in terms of attractiveness to CSP investments, as of mid-2013.

   In addition to geographic and climatic advantages, the state of Rajasthan claimed
to have created an enabling environment for solar power plants through policy,
infrastructure, and facilitation (Pandey et al. 2012). The plant (and most of India's

---

[3] http://cdm.unfccc.int/Projects/Validation/DB/5S9GKT9EX6GEK4EDHISK0L80T2O5YG/view.
html; pp. 15, 16.

CSP potential, in general) was located to the west, allowing for production to coincide with midday and early evening peak usage in central and eastern India (Ummel 2010) (as all of India is grouped within one time zone).

Observers have summarized the risks succinctly. "Direct normal irradiation (DNI) data is not accurate, financing costs in India are high, banks are hesitant to lend, government support is waning, gas water and land are in short supply. Timelines are too ambitious and margins too low. The resulting cutting of corners has backfired. The complexity of setting up CSP plants has been systematically underestimated" (Englemeier 2013).

Owing to strong price pressure of rival technologies such as solar photovoltaics (PV), CSP had faced almost insurmountable challenges in establishing project viability. In June 2013, German industrial giant Siemens had announced the closure of its CSP business, and the laying off of about 280 personnel, based at its Solel facility in Israel, which the former had purchased for USD 418 million in 2009 (Lee 2013).

A detailed technological assessment of concentrating solar thermal power generation, published in 2010, concluded that based on technical (including thermodynamic), economic, environmental, and qualitative criteria, Fresnel lenses with a secondary compound parabolic collector would best suit power plants installed in northwest India. The study included parabolic troughs (as proposed by Godawari), heliostat fields (plane mirrors focusing solar radiation onto a tower), parabolic dishes, etc., for comparison (Nixon et al. 2010) as shown in Fig. 8.4.

In all, project viability was influenced by no less than twelve independent variables involved in the computation of energy generation cost including solar resource, discount rates, initial capital costs, lifetime of the system, and operational and maintenance expense (Hemandez-Moro and Martinez-Duart 2013). In an

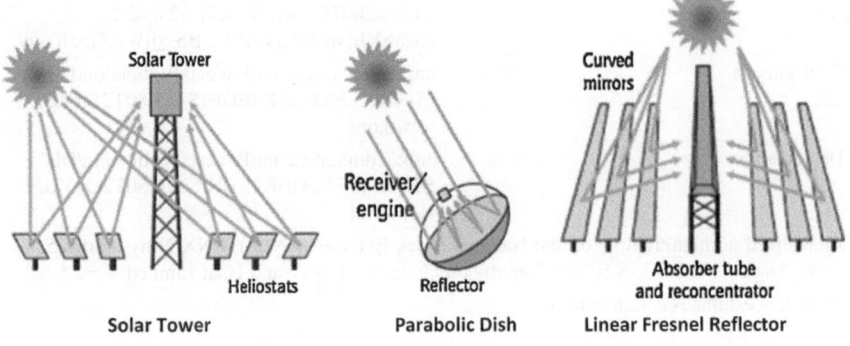

Fig. 8.4 Alternative concentrating solar power technologies

attempt to meet the stringent deadline, Godawari is believed to have spent 20 % in excess of initial estimates of project costs and in return to have generated and sold power sooner than rival plants.

Select Input Data to Estimate Project Returns

Parameter	Unit	Value	Remarks
Load factor of the CSP project	%	29.74	Annual electricity production: Capacity (MW) × PLF (%) × 8760 (h)
Commissioning period	Month	28	Commencing August 20, 2011
The period of assessment	years	25	PPA tenure
Project cost	INR million/ MW	158.8	http://cdm.unfccc.int/Projects/Validation/DB/ 5S9GKT9EX6GEK4EDHISK0L80T2O5YG/ view.html
Electricity tariff	INR/ kWh	12.20	Table 8.1
O&M costs	INR/ MW/a	137,400	
Escalation in O&M at 5.72 % wef March 2011			
Equity	%	21.60 %	http://cdm.unfccc.int/filestorage/1/Q/E/ 1QESIRK5GW82PA6VFT9JLNX3M07HOY/ Untitled%20%28uploaded%2015%20Feb% 2012%2007%3A13%3A31%29.pdf?t= a0d8bXJrNmM5fDBZiBUB6s_jiWoXI5cJOjv9
Debt	%	78.40 %	http://cdm.unfccc.int/filestorage/1/Q/E/ 1QESIRK5GW82PA6VFT9JLNX3M07HOY/ Untitled%20%28uploaded%2015%20Feb% 2012%2007%3A13%3A31%29.pdf?t= a0d8bXJrNmM5fDBZiBUB6s_jiWoXI5cJOjv9
Debt interest rate	%	12.25 %	http://cdm.unfccc.int/Projects/Validation/DB/ 5S9GKT9EX6GEK4EDHISK0L80T2O5YG/ view.html
Depreciation	%	5.28 %	http://cdm.unfccc.int/Projects/Validation/DB/ 5S9GKT9EX6GEK4EDHISK0L80T2O5YG/ view.html

Annualized nominal returns on the National Stock Exchange (NSE), CNX Nifty April 25, 2006–August 14, 2013: 8.50 %[a]; Returns on Godawari Power and Ispat Limited = −2.5 %[a]

[a] Author's estimates/computations

Teaching Note

Case Synopsis

The north-western Indian states of Rajasthan and Gujarat and adjoining parts of the central Indian state of Madhya Pradesh are endowed with copious amounts of solar radiation and have attracted solar energy project investors in droves. However, several projects had suffered setbacks on account of larger-than-expected error margins in available solar radiation data, inadequacies in support infrastructure including the grid network for evacuating the power generated and water supply where required. More importantly, the adverse impact of dust storms on solar photovoltaic as well as solar thermal installations and sand blasting and pitting of reflecting surfaces of concentrating solar power (thermal) plants had been over-looked or underestimated at the very least.

Among the seven concentrating solar power plants (CSP) awarded during the first round of the Indian Government's National Solar Mission, five were to be located in Rajasthan and one each in Gujarat and Andhra Pradesh. The 50 MW plant constructed by Godawari Green Energy Limited was commissioned in June 2013, delayed by one month, and at a cost overrun of about 20 % over initial estimates. Each of the other plants was delayed by many months, with work on a few yet to be even initiated. Considering that the project developers had faced genuine difficulties, the Ministry of New and Renewable Energy (MNRE), Government of India (GoI), extended the project implementation window by ten months, i.e., through March 2014. In order to keep up its contract with the Ministry, *Godawari* had paid premium prices for imported components and had crashed the project to ensure timely completion. This case calls for a detailed analysis of the crashing strategy and its returns to the investors.

Case Question

Analysts are required to compute the benefits and costs from *Godawari's* project crashing (time-cost trade-off) strategy.

Teaching Objectives

- Enabling students to extract relevant data from multiple sources to prepare cash flow estimates for a stand-alone project.
- Preparing a financial model for the project with the inputs available.

- Preparing a cash flow statement for each alternative scenario: the actual (cra-shed) scenario with 20 % higher project costs and 10 months of power gener-ation and a delayed scenario with generation commencing April 2014.
- Analyzing the benefits and costs of the company's project crashing strategy.

Case Objectives and Use

The case calls for organizing a robust financial model and statement of cash flows for each of the alternative scenarios to study the impact of differences in timing of cash flows. The inputs to the financial model are modified to provide for the development of the time-cost trade-off scenarios.

In the present case, the project is implemented within a window of 29 months—1 month behind the 28-month schedule—but at an additional capital expenditure of 20 % over initial estimates. The alternative scenario would have been to stretch implementation to cover 38 months but in the bargain, forfeit the revenues from about 9 months' worth of power generation. Returns on equity investments are computed and compared to the landed cost of debt. The additional variable, con-sequent to the delayed commissioning, could be the potential liquidation of the company's guarantee by the Ministry, as originally contracted.

The case provides for a hands-on managerial decision-making situation where suppliers seek to exploit the time constraints imposed by the project contract. Managers would be required to balance between the additional investments into timely completion and the costs of not doing so. Yet six of the seven conces-sionaires appear to have preferred to delay implementation. This approach provides an interesting contrast to *Godawari's* project implementation strategy.

Teaching Plan

The case presents a project implementation strategy wherein the agency concerned was required to strike a balance between timely commissioning of the project to garner the economic and goodwill benefits, on the one hand, against the additional costs involved, especially from higher prices of imported components and niche products, on the other. Equity investors are last in the pecking order of claimants and yet, given the realization from project operations, in this case, the risk premium received by the equity investors might be marginal, if at all. In the event of default on debt service, lenders gain management control of the project and enforce their security interests. The instructor could guide the course participants to analyze the financial model for two alternative scenarios of expedited completion and delayed commissioning and to weigh the results against the prospect of forfeiting the caution deposit provided. The instructor could then guide the discussion on enhancing returns to equity holders to compensate for the additional risk borne by modifying debt service patterns, among other things.

The instructor could guide the students to evaluate the project in a number of small steps as laid out here in below:

1. Table 1—Replicating the net cash flow model as the project was actually executed.
2. Table 2—Adjusting the net cash flow model presented for alternative project cost estimates and time input data.
3. Analyze the loss or gain subject to the liquidation of the bank guarantee by the Ministry.

What Happened Next

The company published its annual report for fiscal year April 2012–March 2013 in May 2013, reporting total assets of INR 775.25 crore (INR 7752.589 million), and with income from operations yet to accrue as of March end. The balance sheet and profit and loss accounts as reported are appended to the case. The National Renewable Energy Laboratory (NREL), a national laboratory of the US Department of Energy, published the summary of the operational *Godawari* CSP project on the seventh of June, 2013,[4] mentioning among other things that the project was not equipped with thermal storage. The Rajasthan Renewable Energy Corporation Limited had issued the "commissioning certificate" on July 2, and the company notified the exchanges of the project's commissioning, forthwith.

The cash flow chart for the project would need to be drawn up one step at a time:

1. Input data/replication of base case

Capacity	MW	50
Project first cost	INR	9,528,000,000
Debt/equity	%	78.40:21.60
PLF (efficiency factor)	%	29.74
Hours per year	no.	8,760
Model tariff	INR/kWh	12.20
O&M	INR/MW/annum	137,400
Escalation	% per annum	5.72 %

---

[4] http://www.nrel.gov/csp/solarpaces/project_detail.cfm/projectID=247, last accessed 20 May 2014.

Actual scenario	Construction				Operation			
Year	2011	2012	2013	2014[a]	2015	–	2038	2039[b]
	19,939,979	1,356,475,072	6,245,984,949	1,905,600,000		–		
	0.21 %	14.24 %	65.55 %	20.00 %		–		
Revenue from electricity sale				1,208,471,000	1,450,165,200	–	1,450,165,200	241,694,200
O&M expenses				5,725,000	68,70,000	–	24692854.0	4350880.9
Net cash flow	(19,939,979)	(135,6475,072)	(6,245,984,949)	(691,404,000)	1,443,295,200	–	1,425,472,346	237,343,319
Pre-tax return on project	14.36 %					–		

[a] Ten months of project operations in fiscal year 2014
[b] Two months of operations ending May 2038

2. Project funding pattern

	(%)		FYE 2011	FYE 2012	FYE 2013[a]	FYE 2014[a]
Debt	78.40	7,469,952,000			5,564,352,000	1,905,600,000
Equity	21.60	2,058,048,000	19,939,979	1,356,475,072	681,632,949	
Total project cost	100.00	9,528,000,000	19,939,979	1,356,475,072	6,245,984,949	1,905,600,000

[a] Author's estimates

3. Debt amortization schedule

	1	2	3	–	10	11	12
Principal o/s at the beginning of period	7,469,952,000	6,847,456,000	6,224,960,000	–	1,867,488,000	1,244,992,000	622,496,000
Principal repaid during the period	622,496,000	622,496,000	622,496,000	–	622,496,000	622,496,000	622,496,000
Principal o/s at the end of period	6,847,456,000	6,224,960,000	5,602,464,000	–	1,244,992,000	622,496,000	0
Average principal o/s during period	7,158,704,000	6,536,208,000	5,913,712,000	–	1,556,240,000	933,744,000	311,248,000
Interest charge for period	876,941,240	800,685,480	724,429,720	–	190,639,400	114,383,640	38,127,880

Tenure of debt = 12 years; rate of interest = 12.25 % (fixed); and principal repaid in 12 equal installments[a]
[a] Author's estimates

4. Simulated profit and loss account

			FYE March 2013	FYE March 2014	FYE March 2015	FYE March 2016
Revenue from sale of electricity				1,208,471,000	1,450,165,200	1,450,165,200
O&M expenses				5,725,000	6,870,000	7,262,964
EBITDA[a]				*1,202,746,000*	*1,443,295,200*	*1,442,902,236*
Interest charge on project[b]			340,816,560	895,616,120	876,941,240	800,685,480
Depreciation (SLM/ 25 year)				317,600,000	381,120,000	381,120,000
Earnings before tax				*166,163,120*	*185,233,960*	*261,096,756*
Tax charge (33.22 %)				55199388.46	61534721.51	86736342.34
Earnings after tax				*110,963,732*	*123,699,238*	*174,360,414*
Add back depreciation				317,600,000	381,120,000	381,120,000

(continued)

(continued)

			FYE March 2013	FYE March 2014	FYE March 2015	FYE March 2016
Deduct principal repayment					622,496,000	622,496,000
Cash flow available to equity holders	(19,939,979)	(1,356,475,072)	(1,022,449,509)	307,129,880	(114,198,588)	(67,015,586)
Post-tax equity IRR	**11.355 %**					

[a] Project operations commencing June 2013
[b] Interest during construction assuming that principal is o/s for 6 months during FY 2013; o/s principal b/f from FY2013; and o/s principal for FY2014 for 11 months

Cumulative cash flow balance from project operations is positive for each of the years.

5. Input data/counterfactual scenario

Capacity	MW	50
Project first cost	INR	7,940,000,000
Debt/equity	%	78.40:21.60
PLF (efficiency factor)	%	29.74
Hours per year	no.	8,760
Model tariff	INR/kWh	12.20
O&M	INR/MW/annum	137,400
Escalation	% per annum	5.72 %

	Construction				FYE March 2015	FYE March 2039
Counterfactual scenario	19,939,979	1,356,475,072	3,281,792,475	3,281,792,475	—	—
	0.25 %	17.08 %	41.33 %	41.33 %	—	—
Revenue from electricity sale					1,45,01,65,200	1,45,01,65,200
O&M expenses					68,70,000	2,61,05,285
Pre-tax cash flow	(19,939,979)	(1,356,475,072)	142,059,915	(3,281,792,475)	1,443,295,200	1,424,059,915
Pre-tax project return	15.71 %				—	—

6. Project funding pattern

	(%)		FYE 2011	FYE 2012	FYE 2013[a]	FYE 2014[a]
Debt	78.40	6,224,960,000	19,939,979	1,356,475,072	2,943,167,526	3,281,792,475
Equity	21.60	1,715,040,000	19,939,979	1,356,475,072	338,624,949	3,281,792,475
Total project cost	100.00	7,940,000,000	19,939,979	1,356,475,072	3,281,792,475	3,281,792,475

[a] Author's estimates

7. Debt amortization schedule

	1	2	3	–	10	11	12
Principal o/s at the beginning of period	622,496,0000	5,706,213,333	5,187,466,667	–	1,556,240,000	1,037,493,333	518746666.7
Principal repaid during the period	518,746,667	518,746,667	518,746,667	–	518,746,667	518,746,667	518,746,667
Principal o/s at the end of period	5,706,213,333	5,187,466,667	4,668,720,000	–	1,037,493,333	518746666.7	0
Average Principal o/s during period	5,965,586,667	5,446,840,000	4,928,093,333	–	1,296,866,667	778,120,000	259373333.3
Interest charge for period	730784366.7	667,237,900	603691433.3	–	158866166.7	95,319,700	31773233.33

Tenure of debt = 12 years; rate of interest = 12.25 % (fixed); and principal repaid in 12 equal installments[a]

[a] Author's estimates

8. Simulated profit and loss account

			FYE March 2013	FYE March 2014	FYE March 2015	FYE March 2016
Revenue from sale of electricity					1,450,165,200	1,450,165,200
O&M expenses					68,70,000	7,262,964
EBITDA[a]					1,443,295,200	1,442,902,236
Interest charge on project[b]			180,269,011	561,547,811	730,784,367	667,237,900
Depreciation (SLM /25 year)					317,600,000	317,600,000
Earnings before tax					394,910,833	458,064,336
Tax charge (33.22 %)					131189378.8	152168972.4
Earnings after tax					263,721,455	305,895,364
Add back depreciation						
Deduct principal repayment						
Cash flow available to equity holders	(19,939,979)	(1,356,475,072)	(518,893,960)	(561,547,811)	62,574,788	104,748,697
Post-tax equity IRR	12.950 %					

[a] Project operations commencing April 2014

[b] Assuming that principal is o/s for 6 months during FY 2013; o/s principal b/f from FY2013; and o/s principal for FY2014 for 6 months

9. Simulated profit and loss account

			FYE March 2013	FYE March 2014	FYE March 2015	FYE March 2016
Revenue from sale of electricity					1,45,01,65,200	1,45,01,65,200
O&M expenses					68,70,000	7262964
EBITDA^a					1,443,295,200	1,442,902,236
Interest charge on project*			18,02,69,011	56,15,47,811	730,784,367	667,237,900
Depreciation (SLM /25 year)					31,76,00,000	31,76,00,000
Earnings before tax					*39,49,10,833*	*45,80,64,336*
Tax charge (33.22 %)					131189378.8	152168972.4
Earnings after tax					*26,37,21,455*	*30,58,95,364*
Add back depreciation						
Deduct principal repayment						
Cash flow available to equity holders	*(1,99,39,979)*	*(1,35,64,75,072)*	*(51,88,93,960)*	*(1,088,547,811)*	*6,25,74,788*	*10,47,48,697*
Post-tax equity IRR	**11.597 %**					

Alternate scenario: The Ministry decides on liquidating the company's default guarantee amount of INR 527 million in March 2014

Appendix: Balance Sheet and Profit and Loss Accounts for Godawari Green Energy Limited for Years Ended March 2011–2013

Godawari Green Energy Limited			
Balance sheet as at 31 March			
	2011	2012	2013
Equity and liabilities			
Shareholders' funds			
Share capital	121500000.00	121500000.00	1081500000.00
Reserves and surplus	1089142437.00	1090440543.00	1090816594.00
	1210642437.00	**1211940543.00**	**2172316594.00**
Non-current liabilities			
Long-term borrowing	Nil	459942427.00	3220400048.00
Current liabilities			
Short-term borrowing	610000000.00	618399595.00	2262417631.00
Other current liabilities	8831048.00	12188899.00	97455265.00
	618831048.00	**1090530921.00**	**2359872896.00**
Total	*1829473485.00*	*2302471464.00*	*7752589538.00*
Assets			
Non-current assets			
Fixed assets			
Tangible assets	36746.00	510142874.00	516532151
Capital work in progress including preoperative expenses	19903233.00	865826177.00	7059457734
Long-term loans and advances	Nil	446000.00	546000.00
	19939979.00	**1376415051.00**	**7576535885.00**
Current assets			
Cash and bank balances	1173534491.00	377799024.00	155276063
Short-term loans and advances	617267252	544876702.00	10443123.00
Other current assets	18731763.00	3380687.00	10334467
	1809533506.00	926056413.00	176053653.00
Total	*1829473485.00*	*2302471464.00*	*7752589538.00*
Godawari Green Energy Limited			
Profit and loss account for the year ended March			
	2011	2012	2013
Income			
Interest on fixed deposits	26378822.09	54468570.00	16508949.00
(TDS on interest: Rs. 2637882)			
Expenses			
Salary	150000.00	9828692.00	6207340.00
Other expenses		35415099.00	8317214.00

(continued)

(continued)

Godawari Green Energy Limited			
Profit and loss account for the year ended March			
	2011	2012	2013
Audit fee	22060.00		
Communication	8686.00		
Miscellaneous expenses	33191.00		
Finance costs	19386617.32	7087048.00	162200.00
Printing and stationery	74893.00		
Legal fees	1714250.00		
Membership fees	15000.00		
Traveling expenses	3863403.00		
Depreciation	6154.00	517890.00	1357148.00
Preliminary, project, and preoperative expenses written off	922131.00		
Total	**26196385.32**	**52848729.00**	**16043902.00**
Profit before tax	*182436.77*	*1619841.00*	*465047.00*
Provision for tax	40,000	321,734	88,996
Profit after tax	*142436.77*	*1298107.00*	*376051.00*

References

Ahmed F (2010) Energy: mission improbable. Businessworld, 1 Feb 2010, p 44

Englemeier T (2013) Whatever is happening to CSP in India. http://bridgetoindia.com/blog/?p=1504. Accessed 14 Aug 2013

Hemandez-Moro J, Martinez-Duart JM (2013) analytical model for solar PV and CSP electricity costs: present LCOE values and their future evolution. Renew Sustain Energy Rev 20:119–132

Hitchin P (2013) What are the future prospects for utility-scale CSP? Power Engineering International, pp 72–77

India Solar Weekly Market Update (2013) Penalties for CSP projects likely to be deferred by 10 months as no project is ready for timely commissioning. http://bridgetoindia.com/blog/?p=1628. Accessed 14 Aug 2013

Lee A (2013) Siemens to shut solar operation. REChargeNews.com. http://www.rechargenews.com/solar/europe_africa/article1330137.ece. Accessed 14 Aug 2013

Muirhead J (2013) Value of CSP becomes increasingly apparent in SA. http://social.csptoday.com

Nixon JD, Dey PK, Davies PA (2010) Which is the best solar thermal collection technology for electricity generation in north-west India? Evaluation of options using the analytical hierarchy process. Energy 35:5230–5240

Paliwal A (2013) Sun block. Down To Earth, 15 May 2013. http://www.downtoearth.org.in/content/sun-block. Accessed 14 Aug 2013

Pandey S, Singh VS, Gangwar NP, Vijayvergia MM, Prakash C, Pandey DN (2012) Determinants of success for promoting solar energy in Rajasthan, India. Renew Sustain Energy Rev 16(6):3593–3598

Pearson NO (2013a) Asia's biggest CSP plant sells power at half price in India. Bloomberg, 17 June 2013. http://www.renewableenergyworld.com/rea/news/article/2013/06/asias-biggest-csp-plant-sells-power-at-half-price-in-india. Accessed 14 Aug 2013

Pearson NO (2013b) India may defer fines on $1 billion solar thermal plants. Bloomberg.com, 8 May 2013. http://www.bloomberg.com/news/2013-05-08/india-may-defer-fines-on-1-billion-solar-thermal-plants.html. Accessed 14 Aug 2013

Purohit I, Purohit P (2010) Techno-economic evaluation of concentrating solar power generation in India. Energy Policy 38(6):3015—3029

Ross K (2013) Renewables investment: where to get the biggest bang for your green buck. Power Engineering International, pp 14–16

Singh R (2013) Solar power projects miss deadline. The Times of India, Jaipur edn, 20 July 2013. http://timesofindia.indiatimes.com/city/jaipur/Solar-power-projects-miss-deadline/articleshow/21194163.cms?referral=PM. Accessed 20 May 2014

Subramaniam K (2010) Solar power: cutting through the clutter. Businessworld, 15 Mar 2010, p 18

Ummel K (2010) Concentrating solar power in China and India: a spatial analysis of technical potential and the cost of deployment. Working paper no. 219, Center for Global Development, Washington DC

Chapter 9
The Tesla Roadster: Running on Empty?

Valuing Distant Cash Flows and Real Options

It's a head-turner, jaw-dropper...If this is the future of the automobile, I want it.
—Warren Brown, The Washington Post, 25 January 2009

Tesla is arguably making the most exciting car in the world today.
—Gregory Kats, President of Capital-E and
Fmr. Director of Financing for RE and EE,
Department of Energy, William J. Clinton Administration,
International Herald Tribune, 24 May 2013

Background

On Wednesday, May 22, 2013, electric car maker Tesla Motors[1] repaid a USD 465 million loan that the US Department of Energy (DoE) had made to it in the year 2010. Cofounder and chief executive Elon Musk thanked the Energy Department, Congress, and the tax payers for the loan. "I hope we did you proud", he said on the occasion. Tesla was hailed the lone success story in the EV market (Fig. 9.1).

The milestone-based DoE loan made under the Advanced Technology Vehicle Manufacturing Program was used to build two manufacturing units in California, one each, for the battery—motor—electrical equipment assembly, and for the "Model S" Sedan (Lacey 2013). On the strength of such infrastructure, Tesla produced and sold a total of 5,000 electric vehicle (EV) units in Q1—2013 at a 17 % gross margin—including zero-emission-credits—and 5 % on their own merit. The company expected to sell a total of 21,000 units in calendar year 2013.

To overcome "range anxiety," and to give EV owners the "freedom to travel," the company had also proposed to expand its off-grid, (solar-powered), *supercharger,*

---

[1] Tesla Motors, Inc. (NASDAQ GS-"TSLA"); the company was named after inventor and engineer Nikola Tesla who developed the three-phase electric motor like the one used in the Tesla Roadster.

© Springer India 2015
S. Sunderasan, *Cleaner-Energy Investments*,
DOI 10.1007/978-81-322-2062-6_9

Fig. 9.1 The Tesla logo

and fast-charging network and to build 200 such stations over 2014–2015, at distances of 80–100 miles of each other on major routes throughout the USA and Canada (Wesoff 2013).

Company History and Lead to the Case Question

Martin Eberhard and Marc Tarpenning had founded a company based on a portable e-book reader. Encouraged by the overwhelming success of the electronic gadget, they decided that they could create an electric car that had mass appeal. Elon Musk, cofounder of the Internet payment system, *PayPal* contributed USD 30 million in investment funds and joined the company as CEO and product architect. Musk had claimed that he had started Tesla with the goal of mass marketing an affordable electric car and that he would stand by the company until this goal was reached (Postelwait 2013).[2]

The newly founded company chose to partner with England/Hethel-based *Lotus* for modifying one of the existing designs—the *Lotus Elise*—and for producing cars in small runs against confirmed orders.

On May 15, 2013, the company announced an offer of 2.7 million shares of common stock and USD 450 million worth of convertible senior notes due in the year 2018.[3] In addition to these concurrent offers, Tesla had granted the underwriters an option to purchase an additional 0.4 million shares of common stock and USD 67.5 million in aggregate principal amount of notes. It was estimated that the

---

[2] Elon Musk was a quintessential technology professional, first becoming famous when he sold his online payments venture *PayPal* to *eBay* for USD 1.5 billion, and then with his space travel company *SpaceX* becoming the first privately funded company to send a cargo payload to the International Space Station and lately for the success achieved by his renewable energy service company SolarCity. *Ib id.*

[3] Elon Musk himself acquired USD 100 million worth of common equity at the time of the company's concurrent offering.

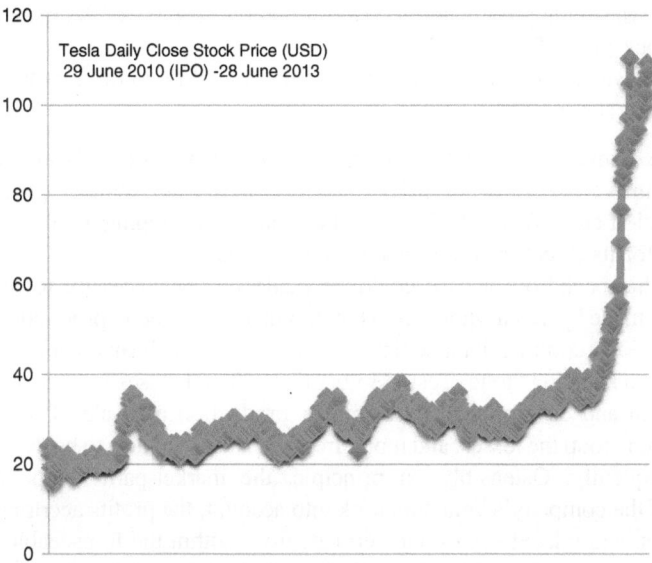

Fig. 9.2 Tesla Motors' daily closing stock prices: June 2010–June 2013: figure developed by author

gross proceeds from the offer would exceed USD 830 million and could even have reached USD 1 billion, which Tesla intended to use to repay the DoE loan.

The evolution in Tesla's stock prices between the initial public offering (IPO) made in June 2010 and June 28, 2013, is shown in the Fig. 9.2. Between the 8th of May 2013, when the Tesla stock closed at USD 55.79 a share and the 28th of May when the stock peaked at 110.33, its value had doubled over a mere 14 trading sessions.[4]

During Q1-2013, Tesla had received USD 68 million in "zero-emission-credits": This stream of cash inflows was expected to cease by 2013. The company's income statement, balance sheet, and cash flow statements for the year 2012, filed with the US Securities and Exchange Commission (SEC) on March 7, 2013, are annexed to this case. Analysts had observed that at USD 100 per share, Tesla's market capitalization amounted to about 25 % of General Motors' (GM) market valuation and had raised concerns on whether the rapid rise in valuation was a result of irrational exuberance, with flavors of the Internet fueled bubble of the late 1990s (Taylor 2013).

A few aspects of the repayment stand out in this context:

(i) Tesla repaid the loan nine years in advance,
(ii) not aided by vehicle sales,

---

[4] Based on daily closing stock prices published on http://finance.yahoo.com; and http://finance.yahoo.com/news/tesla-motors-announces-offerings-common-202535246.html.

(iii) but by raising approximately USD 1 billion on the back of the meteoric rise in its stock price (Eavis 2013), while

(iv) the repayment cost the company an additional USD 26 m in early-repayment penalty (Hartung 2013).

Yet, the company's dependence on policy support did not really end, as Tesla's cars continued to receive tax credits and revenues from selling "green credits" (zero-emission-credits) to other automakers, and was hoping to receive special approvals for its direct-from-the-factory, online sales model.

Given this backdrop, a case could be made for the necessity to analyze the investment made by Tesla Motors into the design, development, production, and sale of one of its early commercial model the Tesla *Roadster*. In the manner of viewing the *Roadster* as a stepping stone, Tesla Motors had suffered losses from its investments into research and development and from its production and sale of the *Roadster,* expecting to recoup the losses, and more, from the sale of models to be developed and sold subsequently. Ostensibly, in principle, the market-participants' consensus estimate of the company's valuation took into account, the profits accruing from the sale of EVs, and related spares and services from within the foreseeable future. In addition, the company would necessarily have had to recover the investments made into the *Roadster* project. Analysts were therefore required to assess the value of the "option to launch newer models," offered by the Tesla *Roadster*.

The Tesla Roadster

The Tesla *Roadster* was fast, fancy, handled like a dream and went like a rocket, but was virtually silent and would never burn a single drop of gasoline during its lifetime. The *Roadster* was the world's first high-performance electric car and could accelerate from zero to sixty miles per hour (mph) in about four seconds and cruise at a top speed of about 130 mph; more importantly, it was "highway capable" and could go further than any competing electric car: an estimated 250 miles on a single charge. Given the relative pricing of electricity and gasoline and adjusting for driving styles, a Roadster could cover 150 miles at the price of running a gasoline powered car for just about 110 miles. Figure 9.3 features the author's sketch of the Tesla *Roadster*.

Above all, the Tesla *Roadster* looked and felt like a sports car with its soft-top, satellite navigation system, and electronic communication systems to coordinate with other household gadgets such as motorized garage doors, able to synchronize with the *ipod,* etc., and other such exotic features. It also provided for routine fitments such as heated seats, a stereo with a CD player, ABS brakes, and dual air bags. Given the simplicity of the motor and the transmission, the car had just three gears—two forward gears and one reverse gear. Shifting was manual with no clutch and hence no stalling or "jerk" when shifting gears. Most of the car's parts, systems, and subassemblies were available off-the-shelf, thus shortening design time and reducing start-up cost substantially (Grabianowski 2013). For the first time, people

Fig. 9.3 The Tesla Roadster; author's sketch

believed that an electric car could be practical as well as a glamorous object of desire (Baker 2013).

The energy storage system (ESS) was a battery bank containing 6,831 low-cost, rechargeable lithium-ion cells, whose performance on laptop computers was well established. A computer processor controlled the smooth charge and discharge of each of the 11 sectors, containing 621 cells each. Further, the regenerative braking system captured the kinetic energy from braking and transferred it back into the ESS.[5] The battery management system was extensively "catastrophe tested" to ensure utmost safety: It could isolate the malfunctioning sector(s) and cutoff power supply in case of overheating, abnormal tilting, or sensing of smoke (Sweney 2013).

The first two batches of 100 units each of the limited edition Roadster costing USD 100,000 with a down payment (at the time of booking) of USD 75,000 was sold out very shortly after production commenced in 2008. On account of receiving such advance payments, the company has managed to operate at a very low level of leverage throughout, which was considered consistent with a technology development project.

By mid-2009, the company had made and sold a total of 500 units. At its peak, as of January 2010, the company had made and sold over a 1,000 units of the Roadster across the Americas, Europe, and Asia (Leader 2013). In all, 2,500 of the cars were made and sold over four years ending December 2011 as detailed in Table 9.1.

As with other technology and design-intensive industry sectors, the automobile sector, and more specifically the electric car segment, were rapidly evolving and

---

[5] The energy management technology, the thermal management systems, and the engineering associated with their development could by themselves be adapted for energy storage for utility, community, and residential applications.

Table 9.1 Sales data and financial performance of Tesla Motors

Year (Jan–Dec)	2006	2007	2008	2009	2010	2011
Units sold (numbers)	0	0	200	800	800	700
Gross profit (USD '000)	NO SALE	64	−1,141	9,535	30,731	61,595
Net loss (USD '000)	29,957	78,157	82,782	55,740	154,328	254,411

Source finance.yahoo.com/q/is?s=TSLA+Income+Statement and annual and Form S-1 filed by Tesla Motors, Inc. with the US Securities and Exchange Commission, 29 January 2010

had been intensely competitive. Staying contemporary was a serious challenge for all manufacturers. Slowly, but surely, the appeal of the product had eroded: By end-2011, the Tesla *Roadster* was not the "newest, coolest thing," any longer.

To retain and engage existing customers, *Tesla* designed a buyback scheme to help replace the aging model with a more recent one. The company offered customers the option of swapping the Roadster for a *Model S*, a newer, family sedan costing between USD 57,400 and USD 105,400 offering a range of 300 miles at 55 mph. This was planned as a bridge measure to attract customers back to the *Roadster*, possibly branded the "Model R" scheduled for launch in January 2017.

   i. The pricing of the vehicles turned in under the Tesla Motors buyback scheme would be subject to detailed inspection.
  ii. The company anticipated that some 80 % of the Roadsters would be turned in during calendar 2013, if customers were offered a *weighted average buyback price* of USD 60,000.
 iii. At the two extremes, the company expected that the year 2008 model Roadster driven for some 31,000 miles would resell at about USD 73,300 and a year 2010 or newer model Roadster having done some 2,900 miles for USD 93,500 and the others priced in between.
  iv. The company also hoped to lease the Roadster for USD 1,658 per month, including on-site service and repair.
   v. The on-site repair and service costs are expected to be 8.0 % of lease rental for 2008 and 2009 models and 4.0 % of the lease rental and for 2010 and 2011 models.[6]
  vi. The company believed that 30 % of the turned-in Roadsters would be resold within an average of 12 months from the date of buyback. The remaining 70 % was expected to be leased out for *average* periods ranging from 24 months to 48 months.[7]
 vii. The weighted average cost of capital (WACC) is assumed to be 7.00 % per annum in nominal terms.

The break-even value of the options yielded by the Roadster, to be recovered when the "Model R" would be introduced in January 2017, was to be estimated.

---

[6] Author's estimates for the purpose of the present analysis.

[7] *ib id.*

Tesla Motors, Inc.

Consolidated Balance Sheets (in thousands, except share and per share data)

	December 31, 2012	December 31, 2011
Assets		
Current assets		
Cash and cash equivalents	$ 201,890	$ 255,266
Short-term marketable securities	–	25,061
Restricted cash	19,094	23,476
Accounts receivable	26,842	9,539
Inventory	268,504	50,082
Prepaid expenses and other current assets	8,438	9,414
Total current assets	524,768	372,838
Operating lease vehicles, net	10,071	11,757
Property, plant and equipment, net	552,229	298,414
Restricted cash	5,159	8,068
Other assets	21,963	22,371
Total assets	$ 1,114,190	$ 713,448
Liabilities and stockholders' equity		
Current liabilities		
Accounts payable	$ 303,382	$ 56,141
Accrued liabilities	39,798	32,109
Deferred revenue	1,905	2,345
Capital lease obligations, current portion	4,365	1,067
Reservation payments	138,817	91,761
Long-term debt, current portion	50,841	7,916
Total current liabilities	539,108	191,339
Common stock warrant liability	10,692	8,838
Capital lease obligations, less current portion	9,965	2,830
Deferred revenue, less current portion	3,060	3,146
Long-term debt, less current portion	401,495	268,335
Other long-term liabilities	25,170	14,915
Total liabilities	989,490	489,403
Commitments and contingencies (Note 14)		
Stockholders' equity		
Preferred stock; $0.001 par value; 100,000,000 shares authorized; no shares issued and outstanding	–	–
Common stock; $0.001 par value; 2,000,000,000 shares authorized as of December 31, 2012 and 2011, respectively; 114,214,274 and 104,530,305 shares issued and outstanding as of December 31, 2012 and 2011, respectively	115	104
Additional paid-in capital	1,190,191	893,336
Accumulated other comprehensive loss	–	(3)
Accumulated deficit	(1,065,606)	(669,392)
Total stockholders' equity	124,700	224,045
Total liabilities and stockholders' equity	$ 1,114,190	$ 713,448

Tesla Motors, Inc.

Consolidated Statements of operations (in thousands, except share and per share data)

	Year Ended December 31					
	2012		2011		2010	
Revenues						
Automotive sales	$	385,699	$	148,568	$	97,078
Development services		27,557		55,674		19,666
Total revenues		413,256		204,242		116,744
Cost of revenues						
Automotive sales		371,658		115,482		79,982
Development services		11,531		27,165		6,031
Total cost of revenues		383,189		142,647		86,013
Gross profit		30,067		61,595		30,731
Operating expenses						
Research and development		273,978		208,981		92,996
Selling, general, and administrative		150,372		104,102		84,573
Total operating expenses		424,350		313,083		177,569
Loss from operations		(394,283)		(251,488)		(146,838)
Interest income		288		255		258
Interest expense		(254)		(43)		(992)
Other expense, net		(1,828)		(2,646)		(6,583)
Loss before income taxes		(396,077)		(253,922)		(154,155)
Provision for income taxes		136		489		173
Net loss	$	(396,213)	$	(254,411)	$	(154,328)
Net loss per share of common stock, basic, and diluted	$	(3.69)	$	(2.53)	$	(3.04)
Weighted average shares used in computing net loss per share of common stock, basic, and diluted		107,349,188		100,388,815		50,718,302

Tesla Motors, Inc.						
Consolidated statements of cash flows (in thousands)						
		Year Ended December 31				
		2012		2011		2010
Cash flows from operating activities						
Net loss	$	(396,213)	$	(254,411)	$	(154,328)
Adjustments to reconcile net loss to net cash used in operating activities:						
Depreciation and amortization		28,825		16,919		10,623
Change in fair value of warrant liabilities		1,854		2,750		5,022
Discounts and premiums on short-term marketable securities		56		(112)		–
Stock-based compensation		50,145		29,419		21,156
Excess tax benefits from stock-based compensation		–		–		(74)
Loss on abandonment of fixed assets		1,504		345		8
Inventory write-downs		4,929		1,828		951
Changes in operating assets and liabilities						
Accounts receivable		(17,303)		(2,829)		(3,222)
Inventories and operating lease vehicles		(194,726)		(13,638)		(28,513)
Prepaid expenses and other current assets		1,121		(248)		(4,977)
Other assets		(482)		(288)		(463)
Accounts payable		187,821		19,891		(212)
Accrued liabilities		9,603		10,620		13,345
Deferred development compensation		–		–		(156)
Deferred revenue		(526)		(1,927)		4,801
Reservation payments		47,056		61,006		4,707
Other long-term liabilities		10,255		2,641		3,515
Net cash used in operating activities		(266,081)		(128,034)		(127,817)
Cash flows from investing activities						
Purchases of marketable securities		(14,992)		(64,952)		–
Maturities of short-term marketable securities		40,000		40,000		–
Payments related to acquisition of Fremont manufacturing facility and related assets		–		–		(65,210)
Purchases of property and equipment excluding capital leases		(239,228)		(184,226)		(40,203)
Withdrawals out of (transfers into) our dedicated Department of Energy account, net		8,620		50,121		(73,597)
Increase in other restricted cash		(1,330)		(3,201)		(1,287)
Net cash used in investing activities		(206,930)		(162,258)		(180,297)

(continued)

(continued)

Tesla Motors, Inc.

Consolidated statements of cash flows (in thousands)

	Year Ended December 31		
	2012	2011	2010
Cash flows from financing activities			
Proceeds from issuance of common stock in public offerings, net	221,496	172,410	188,842
Proceeds from issuance of common stock in private placements	–	59,058	80,000
Principle payments on capital leases and other debt	(2,832)	(416)	(315)
Proceeds from long-term debt and other long-term liabilities	188,796	204,423	71,828
Principle payments on long-term debt	(12,710)	–	–
Proceeds from exercise of stock options and other stock issuances	24,885	10,525	1,350
Excess tax benefits from stock-based compensation	–	–	74
Deferred common stock and loan facility issuance costs	–	–	(3,734)
Net cash provided by financing activities	419,635	446,000	338,045
Net increase (decrease) in cash and cash equivalents	(53,376)	155,708	29,931
Cash and cash equivalents at beginning of period	255,266	99,558	69,627
Cash and cash equivalents at end of period	$ 201,890	$ 255,266	$ 99,558
Supplemental disclosures			
Interest paid	$ 6,938	$ 3,472	$ 1,138
Income taxes paid	117	282	9
Supplemental non-cash investing and financing activities			
Conversion of preferred stock to common stock	–	–	319,225
Issuance of common stock upon net exercise of warrants	–	–	6,962
Issuance of convertible preferred stock warrant	–	–	6,294
Issuance of common stock warrant	–	–	1,701
Acquisition of property and equipment	44,890	15,592	4,482

Teaching Note

Case Synopsis

Technology-intensive product development involves huge up-front investments into research design and development, prototype testing, standardization, and eventual commercialization. Often, such start-up ventures spend several years before they manage to achieve critical mass in sales volumes and to break even on their initial investments. Yet, such ventures are of great interest to equity investors owing to the potential upside that a breakthrough could generate, forcing well-established technology and mature industry players to play catch-up. The *Roadster* was the first of several EVs proposed to be commercially launched by California/ Palo Alto-based Tesla Motors. The company had made substantial investments into the design and development of the product and a further few years making losses from the sale of the *Roadster*.

The case provides a brief background relating to the formation and commercial performance during the early years of Tesla's existence. The financial statements presented in the annex and data summarized within the body of the case provide a perspective on the cost structures and profitability, or otherwise, of the company's operations. Management had chosen to build and sell the *Roadster* incurring losses along the way. Investors seemed to perceive that in the years to come, the company would recoup all the investments and losses to date, and more, which perception was reflected in the heightened valuation of the firm. The company expected to launch the renewed *Roadster*, possibly branded the "Model R," in January 2017.

By building and selling the *Roadster* and incurring business losses in the process, the company has managed to secure brand recognition and valuable experience with customers and with the technology itself. Presumably, the learning was to translate into superior and cost-effective products in the near-term. The case is intended to value the *Roadster* as a stand-alone project and to assess the break-even value of the option that it had provided the company with.

Case Question

What would the value of the option-to-launch future models provided by the Roadster be, as assessed at the beginning of year 2017, when the "Model R" was scheduled to be launched and discounted to end-December 2012, when *Roadster* production was suspended.

Teaching Objectives

- Enabling students to extract relevant data to prepare cash flow estimates for a stand-alone project within a multi-product company.
- Providing students with the experience of undertaking a benefit-cost analysis of a real-world situation and to compare values across different points in time by compounding/discounting.
- Enable students to alter input values and assess the sensitivity of key variables.
- For students to compare market valuation of the firm with their own assessments of the break-even valuation and thereby to compute the premium placed on the firm.

Case Objectives and Use

The case deals with the analysis of real options that most companies are faced with, but a situation in which few managers would initiate multi-year loss-making projects. Managers who do assume such risks do so hoping that such a project would provide the firm with a firm foothold in the market. Through the same product or through subsequent models, through sales of spares or consumables or through service provision, such investments would need to be recovered. In the event that the losses from the development projects and the pilot commercial product are not recovered, such investments would have to be amortized and written-off and would result in a net destruction of shareholder wealth. For management to be working in the interest of maximizing shareholder wealth, the introductory offer made at a loss should be part of a longer term plan which demonstrated that such investments would be recouped with reasonable returns, meeting shareholder expectations. This case is one such exercise where the *Roadster* (2008–2011) was presumed to provide the option-to-launch subsequent models, and course participants are required to assess the value that the company would need to generate from subsequent activities (2013–2017) for the launch to breakeven—over and above the returns on these activities themselves.

The instructor could choose to expand the scope of the case to include other aspects that could impact the decision to launch a loss-making principal product and whether the losses could be made up by complementary products viz., battery packs, charging station infrastructure etc., and on whether Tesla could engage in some form of complementary product (razor-and-blade) pricing strategies. In other words, the instructor could hint at investing in a present loss-making venture such as the technology-demonstrating *Roadster*, a buyback, and a subsequent lease of the same product, with the prospect that the investments could be recovered by secondary activities viz., newer models, higher margins on the "Model R," etc.

Teaching Plan

The case presents a scenario where a market segment needed to be carved out from an existing and well-entrenched market that was dominated by large global players. The product slated to be introduced was more expensive compared to conventional alternatives and was expected to sell in smaller numbers. This was apparently an impossible situation, as has been demonstrated by the collapse of several rivals viz., Fisker, Coda, Bright Automotive, etc., and by the suspension of, or delays in, EV production by several "mainstream" auto majors. The success of the strategy hinged on the promise that the product would entice consumers and that they would return to buy newer launches at ever-increasing margins to cover for the losses suffered in the early years. The instructor could extend the discussion to cover various risks involved in Tesla's financing plan and the risk management measures, including replacing debt with equity, the launch of early models expecting to recoup research-related investments from subsequent models, etc.

The instructor could guide the students to evaluate break-even value in a number of small steps as laid out herein below:

1. Table 1—Buyback data: preparing a table listing the numbers of units surrendered corresponding to each year of production.
2. Table 2—Resale data: preparing a table estimating the revenues from the resale of the surrendered units corresponding to each year of production.
3. Table 3—Lease data: preparing a table estimating the revenues from the lease of the surrendered units for the respective year of production, assuming that all units are leased for the average lease period.
4. Computing the gross margin earned by the company if it were to offer a weighted average buyback price of USD 60,000? Conversely, the buyback price could be computed for a desired gross margin.

Tesla expected to launch the "Model R" a rehashed and upgraded version of the *Roadster* by January 2017. The Company might have taken the view that the product development and marketing expenditure incurred, and hence, the losses suffered during production and sale and from the buyback (if any), had given it the [real] option to launch the newer and higher margin earning product.

5. In principle, analysts would be required to compute the threshold real option *contribution* value that the "Model R" needed to demonstrate as of January 2017, to justify the *Roadster*-related investments [over and above the value required to justify Model R's/other models' own design, development, production, and sale].

YEAR (Jan–Dec)	2006	2007	2008	2009	2010	2011
Units sold	0	0	200	800	800	700
Gross profit	NO SALE	64	−1,141	9,535	30,731	61,595
Net loss ('000 USD)	29,957	78,157	82,782	55,740	154,328	254,411

		No. of years	Compounded to end-2012
		6	44,957,379
		5	109,619,236
		4	108,510,315
		3	68,283,897
		2	176,690,127
		1	272,219,770
Loss from operations and sale			78,02,80,724
Loss from buyback, lease, and resale			56,96,651
Total loss as of Jan 2013			*78,59,77,375*
Total loss as of Jan 2017			***1,03,02,56,007***

Given the assumptions and estimates, the break-even value that needs to be recouped to justify the launch, sale, buyback, and lease/resale of the Roadster would exceed USD 1 billion as of January 2017.

What Happened Next

On July 1, 2013, two rather contradicting opinions were voiced. Eric Schaal wrote on the Wall St Cheet Sheet,[8] listing seven affordable EVs built for city use. While Tesla was credited with changing the way people viewed electric cars, the company was listed number seven on the list, but without a specified affordable EV yet. On the very same day, the Tesla stock gained almost 10 % and closed at USD 117.18 a share, apparently following a report by Elaine Kwei, Business Analyst at Jeffries, who pegged the stock at USD 130 a share based on a 10 year discounted cash flow forecast.[9] In May 2014, finding Tesla Motors "vulnerable" owing to its small scale relative to its peers in the automobile industry, and its narrow product focus, ratings agency Standards and Poor's (S&P) rated Tesla's bonds "B-" ("Junk", six levels below investment grade): The rating was unsolicited and the company responded saying that the agency had not received feedback on their growth plans.[10]

---

[8] http://wallstcheatsheet.com/stocks/7-electric-cars-made-for-the-city-and-a-budget.html/?a=viewall, last accessed 30 July 2013.

[9] http://finance.yahoo.com/news/teslas-stock-surged-9-bullish-201900398.html, last accessed 30 July 2013

[10] http://www.forbes.com/sites/steveschaefer/2014/05/28/sp-gives-tesla-a-junk-bond-rating-cites-vulnerable-business/, and http://www.bloomberg.com/news/2014-05-27/tesla-gets-unsolicited-junk-grade-from-s-p-on-niche-position.html, last accessed 15 June 2014.

The cash flow chart for the project would need to be drawn-up one step at a time:

1. Roadster buyback data

					Total
Make year	2008	2009	2010	2011	
Units made	200	800	800	700	2,500
% Units turned-in	80	80	80	80	
Numbers of units	160	640	640	560	2,000

2. Roadster resale data

					Total
% Numbers sold	30		30	30	30	
Numbers sold	48		192	192	168	600
Selling price	73,300		83,400	93,500	93,500	
Revenue from sale	3,518,400		16,012,800	17,952,000	15,708,000	53,191,200 (as of mid-2014)
Total revenue discounted to mid-2013					**4,97,11,402**

3. Roadster lease data

Lease price per month		1,658	USD
Min lease period		24	months
Max lease period		48	months
Average lease		36	months

4. Roadster lease revenue, adjusted for timing of cash flow

					Total
% Numbers leased	70	70	70	70	
Numbers leased	112	448	448	392	1,400
Lease rental per month	1,658	1,658	1,658	1,658	
Service repair (%)	8	8	4	4	
Revenue from lease per month	1,525.36	1,525.36	1,591.68	1,591.68	
Average lease tenure	36	36	36	36	
	6,150,252	24,601,006	25,670,615	22,461,788	78,883,661 (as of mid-2016)
Total revenue discounted to mid-2013					**6,43,92,565**
Weighted average buyback price (to achieve 80 % surrender rate)					**60,000.00**

Buyback price	12,00,00,000
Total revenue	11,41,03,967
Gross margin	−4.9134 %

5. Roadster loss from sale, buyback, lease, and resale, adjusted for timing of cash flow

LOSS	Loss discounted by 6 months
58,96,033	56,96,651
(as of mid-2013)	(as of beginning of 2013)

References

Baker DR (2013) Tesla starts roadster buyback program San Francisco Chronicle, 17 Oct 2012, Hearst Communications Inc. http://www.sfgate.com/business/article/Tesla-starts-Roadster-buyback-program-3957536.php#ixzz2JNtyMma0. Accessed 1 July 2013

Eavis P (2013) Electric car maker repays U.S. clean-energy loan early. International Herald Tribune

Grabianowski E (2013) How the Tesla roadster works, www.howstuffworks.com. http://auto.howstuffworks.com/tesla-roadster1.htm. Accessed 28 Jan 2013

Hartung A (2013) Why Tesla is beating GM, Nissan and Ford. http://www.forbes.com/sites/adamhartung/2013/06/28/why-tesla-is-beating-gm-nissan-and-ford/. Accessed 1 July 2013

Lacey S (2013) Tesla motors repays its government loan in full. Greentech Media. http://www.greentechmedia.com/articles/read/tesla-motors-repays-its-government-loan-in-full?utm_source=Daily&utm_medium=Headline&utm_campaign=GTMDaily. Accessed 1 July 2013

Leader J (2013) Eco-glossary: Tesla roadster. Mother Nature Network. www.mnn.com. Accessed 28 Jan 2013

Postelwait J (2013) Spring time for Tesla. RenewableEnergyWorld.com. http://www.renewableenergyworld.com/rea/blog/post/2013/05/springtime-for-tesla?cmpid=SolarNL-Tuesday-May14-2013. Accessed 28 June 2013

Sweney (2013) Disparities in EV battery philosophy and Tesla's hidden advantage. www.greentechmedia.com. Accessed 21 Jan 2013

Taylor A (2013) 9 questions for Tesla's Elon musk. CNN Money. http://money.cnn.com/2013/06/12/autos/tesla-elon-musk.fortune/index.html. Accessed 30 June 2013

Wesoff E (2013) Tesla super charger network operable during zombie apocalypse. http://www.greentechmedia.com/articles/read/Tesla-Supercharger-Network-Operable-During-Zombie-Apocalypse?utm_source=Daily&utm_medium=Headline&utm_campaign=GTMDaily. Accessed 1 July 2013

Chapter 10
Mount Coffee Hydro: Stimulating A New Generation

Valuing Growth Options

*It was very wonderful to see. It was like a tourist site ...
people would pay a visit, take a picture, and bring their family
to see the view of the dam and the reservoir.*
—Goodrich Zodehgar, Security Supervisor and Electrician

*Nothing was left. Not even a piece of metal. Cables,
conductors, transformers, substations, meters—we literally
lost everything.*
—Shahid Mohammed, CEO Liberia Electricity Corporation

*The restoration of the Mount Coffee Hydro-Electric Plant is
an urgent national priority for Liberia.*
—Joseph N. Boakai, Hon. Vice President of Liberia, October
2010

Background

The 34 MW phase 1 of the Mount Coffee Hydropower Plant in Liberia, western Africa, was commissioned in 1967, with financing from the United States government. Plant capacity was increased to 64 MW in 1973 with financing from the World Bank, with the enlarged project accounting for 35 % of all electricity generated in Liberia (The Liberian Observer 2014). It was then named Walter F. Walker Hydro Dam.

Plans to further double the capacity of the project were abandoned when rebel forces seized the project site in July 1990. The facilities were vandalized and destroyed during the 14-year-long conflict and civil unrest that ended in 2003. Contrasting visuals are depicted in Figs. 10.1 and 10.2.

Electrification levels in Liberia were abysmally low at 3 %, and probably, the lowest rates in the world (compared to a West African average of 28.5 %) and the tariffs (USD 0.43 for grid-supplied power to USD 3.96/kWh for power from diesel generators) were among the highest in the world. This statistic alone reflected the magnitude of Liberia's development challenges. With the return to peace, rehabilitation of the Mount Coffee plant was considered the best available, sustainable, and near-term energy generation option.

© Springer India 2015
S. Sunderasan, *Cleaner-Energy Investments*,
DOI 10.1007/978-81-322-2062-6_10

Fig. 10.1 A view of the project site before commencement of rehabilitation work; *Figure credit* mtcoffeeliberia.com—reproduced with permission

Fig. 10.2 The operational Mount Coffee Hydropower Plant from before the civil war (*left*) and what was left of it (*right*); *Figure credit* mtcoffeeliberia.com—reproduced with permission

In Liberia, energy consumption, economic growth, and employment generation were positively correlated, in that order (Wesseh and Zoumara 2012). Potential delays in renovation and commissioning of the plant were projected to increase consumption of high-cost imported fossil fuels, and to prolong the stagnation of Liberian civil society. The electricity deficit was declared a "national emergency," and it was therefore decided that the renovation project needed to be fast-tracked and be publicly funded. The Project Implementation Unit (PIU), whose logo is displayed in Fig. 10.3, was established within the Liberia Electricity Corporation (LEC), and renovation work was commenced in May 2012.

Target dates for commissioning of the first phase, and the entire project, respectively, were set at 20 December 2015 and 31 December 2016.[1] The upgraded

---

[1] "Overview", Project Implementation Unit, Mount Coffee Hydropower Plant Rehabilitation, http://mtcoffeeliberia.com/?page_id=96, last accessed 27 January 2014.

Fig. 10.3 Logo of the Mount
Coffee Project
Implementation Unit (PIU);
figure credit mtcoffeeliberia.
com—reproduced with
permission

project was of 80 MW in installed capacity and was slated to help transform
Liberia's economy. The ceremonial inauguration of the reconstruction was held on
14 January 2014. Figure courtesy: mtcoffeeliberia.com.

Project Description

Providing "reliable modern energy services" was a key promise by Her Excellency,
President Ellen Johnson Sirleaf, leading to her electoral victory in 2005 making her
the first woman elected head of state in all of Africa.[2] The Liberian National Energy
Policy of 2009 set targets of 30 % urban and 15 % rural electrification rates by the
year 2015. On being reelected president in 2011, she reaffirmed her promise to
"bring back electricity".

Hydropower has been among the oldest sources of energy known to mankind,
making an indispensable contribution to stable electric power supplies and consequently to economic and social development across the globe. The secondary
economic costs of hydroelectric power included the price of land, foregone agricultural output, and resettlement costs. The economic benefits included the avoided
fuel costs, avoided greenhouse gas emissions, improved air quality, conserved
foreign, and currency reserves. According to the International Hydropower
Association, some 30 GW of new hydropower capacity was commissioned in year
2012, including about 2 GW of pumped storage systems, primarily in South
America, Asia, and Africa (Appleyard 2014). In the Liberian context, the Mount

---

[2] The October-1938-born president was also awarded the Nobel Memorial Prize for Peace for the
year 2011 for "non-violent struggle for the safety of women and for women's rights to full
participation in peace-building work".

Coffee Project was also a symbol of political and economic stability and of reduced vulnerability to global oil price shocks.[3] The monthly costs of electricity generated by Mount Coffee was estimated to be within the limits of customers' willingness to pay (at about USD 3.94 per month) (Alfaro and Miller 2014). Since the area was already occupied by an operating project in the past, incremental impacts on flora and fauna in the region and on the human population were slated to me "relatively small" to "negligible".[4]

The Mount Coffee Hydropower Plant (MCHPP) was located in Harrisburg Township, 35 km from the capital city of Monrovia, on the 500-km-long St. Paul River that rose in southeastern Guinea, flowed through Liberia and drained into the Atlantic Ocean. The renovation work at the hydropower plant was projected to cost a total of USD 230 million: this included rehabilitation of the hydropower plant and reservoir, the construction of a 66 kV substation at Mount Coffee, two 66 kV transmission lines between the project site and the capital city of Monrovia, and the expansion of two existing receiving substations in the city. Distribution of electricity to the households was to be funded separately by the Liberia Electricity Corporation. Likewise, the construction of a 225 kV transmission line to export power from the Mount Coffee substation to neighboring countries was to be funded through the West African Power Pool (WAPP).

Means of Financing

Liberia was among the main partners under the Norwegian Clean Energy Initiative and the Government of Norway had played a substantial role in Liberia's energy sector. The Liberia Electricity Corporation itself was administered by Canadian electric power and gas utility, Manitoba Hydro International Limited (www.mhi.ca), under a "Management Contract" running through December 31, 2016. The Project Implementation Unit and the Management Contract were funded by the Norwegian government. Norway had also committed to making a grant of 450 m Norwegian Kroner ($\sim$ USD 75 million). KfW (development finance institution) of Germany had committed to making a grant of €25 million ($\sim$ USD 35 million), and the European Investment Bank (EIB) had provided a concessionary, 20 year/1.4 % interest bearing loan of €50 million ($\sim$ USD 66 million).

The Government of Liberia had provided for annual budgetary support of USD 15 million and was to make up for the balance costs and delays and overruns. In addition, the European Union—Africa Investment Trust Fund (ITF) had made a €1.5 million grant to the West African Power Pool Secretariat to finance the Social Impact Assessment Study and for developing the Resettlement Action Plan

---

[3] Project Benefits, http://mtcoffeeliberia.com/?page_id=33, last accessed 27 January 2014.

[4] http://mtcoffeeliberia.com/wp-content/uploads/2013/03/Final-ESMP-2013-02-08.pdf, last accessed 27 January 2014.

(undertaken by Pöyry Energy Limited, Finland) and for the Engineering design and Preparation of Bidding Documents (undertaken by Stanley Consultants, USA).[5]

In October 2013, German firm Voith Hydro (www.voith.com) was awarded a USD 59.2 million contract to modernize the turbines, to deliver generators, and to control technology and the electro-mechanical power plant equipment. At the end of the January 2014, LEC had commenced work on tenders for construction of the substations, telecommunication facilities, earthing grid, fire protection walls, switchgear, feeders, transmission lines, etc.

The Biggest Was Not Big Enough

- By itself, the renovated Mount Coffee Project was a run-of-the-river project, meaning that it was not backed up by a large reservoir and that it held water for about 2 days' operations.
- The average monthly flow in the river varied from 1,600 m^3/s in September to a mere 108 m^3/sec in March with an annual average of 564 m^3/s. Figure credits: Project Documents, Impact Assessments etc.
- Water throughput required for the power plant to run at full capacity was 412 m^3/s, implying that full capacity production was possible only for 156 days, and for 75 days in a year production, it was possible with 1 of the 4 units only (25 % of installed capacity). No power production was possible for 28 calendar days[6] in a year;
- Consequently, given these limitations, the annual power output from the plant was estimated to be 432 million kWh ("units").

The government of Liberia proposed to enhance power production at the plant and to regulate the seasonal flow by building an upstream reservoir at the confluence of the St. Paul and Via Rivers to ensure round-the-year water availability, Figs. 10.4 and 10.5 (Gaillard 2014). Such storage would increase annual power generation at Mount Coffee by 25 %.

The Via Reservoir added 25 % to Mount Coffee's production, although at a cost. However, Mount Coffee was the lowest in a planned cascade of five dams/power stations and the reservoir made it possible to add to the output of each such intermediate project. In essence, the Via Reservoir would not have justified a potential investment as it was not cost-effective, by itself; in fact, it took value away from the Mount Coffee Project. Its construction was justified on the basis of the value the reservoir added to the intermediate projects: between the Via Reservoir and the Mount Coffee Hydropower Project.

---

[5] http://www.eu-africa-infrastructure-tf.net/activities/grants/mount-coffee-hydropower-plant-rehabilitation.htm, last accessed 27 January 2014.

[6] Environmental and Social Impact Assessment (ESIA) report, p. 203 of 381.

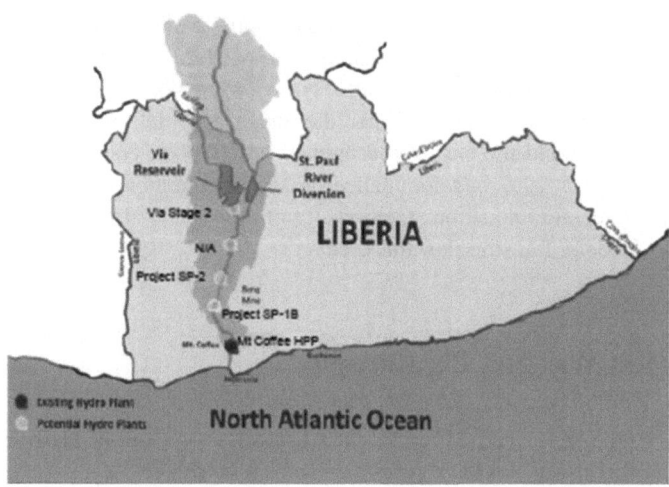

Fig. 10.4 A map showing the relative locations of the Via Reservoir and the Mount Coffee Hydropower Project

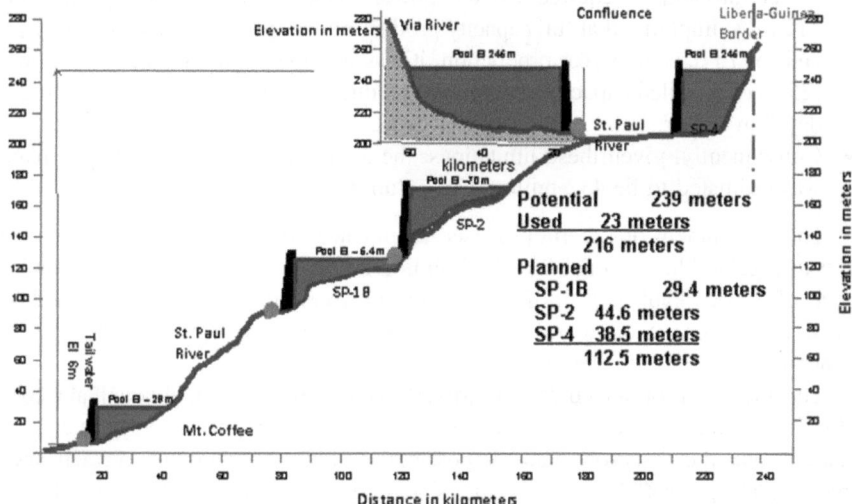

Fig. 10.5 A section through the Via Reservoir, the Mount Coffee Hydropower Project and intermediate hydropower projects; *Source* R.M.Gaillard, "presentation of results" evaluation of EU energy funding in liberia and recommendations, 24 July 2012, MWH, slide 23 of 27, http://eeas. europa.eu/delegations/liberia/documents/press_corner/20120724_01_en.pdf, last accessed 28 January 2014—reproduced with permission

Project Data

Parameter	Unit	Mount Coffee HPP	Mount Coffee HPP + Via Reservoir
Project cost	USD	230,000,000	544,000,000*
Means of finance			
Grants	USD	110,000,000	110,000,000
Debt	USD	50,000,000	200,000,000
(Liberia Electricity Corp) Equity	USD	70,000,000	234,000,000
Cost of debt	% pa	1.4 %	3.0 %
Maximum annual power production	kWh	432,000,000	540,000,000
Production volumes	% of capacity	80 % y1; +5 % each year fm y2; 100 % in y5	80 % y1; +5 % each year fm y2; 100 % in y5
Aggregate technical and commercial losses	%	22 %	22 %
Fixed end-user tariffs (to year 2041)	USD	0.148	0.148
Operation and mainte-nance exp	% of capex	1.00 %	1.50 %
Project life	year	50	50
Tax rate on corporation profits	%	18.30 %	18.30 %
Residual value of pro-ject (in year 8 of ops)	multiple	5 × y7 free cash flow	8 × y7 free cash flow
Discount rate for pub-licly funded projects	% pa	3.5 %	3.5 %

*Author's estimates

Distribution of capex spend

Year	−5	−4	−3	−2	−1	1
Mount Coffee alone		14.60 %	17.00 %	37.90 %	30.10 %	0.40 %
Mount Coffee and Via Reservoir	0.52 %	13.58 %	17.23 %	37.86 %	30.29 %	0.52 %

Teaching Note

Case Synopsis

The Mount Coffee Hydropower Plant was the single largest power generation unit in Liberia. The 14-year civil war between 1990 and 2003 saw the complete destruction of the project. The site was looted and vandalized, leaving the bare structure to narrate the tales of the unfortunate events. With the return to peace, the rebuilding and enlarging the hydropower project was put on high priority.

Given the wide dispersion in seasonality of flow through the St. Paul River on which the power project was located, its output was projected to be sub-optimal, although more cost-effective than diesel generators and fuel oil power plants. The government proposed to construct the Via Reservoir to help regulate flow across seasons and to enhance power output at Mount Coffee by about 25 %. However, the reservoir project by itself was not cost-effective, as it did not have major revenue streams of its own, and the combined project was projected to generate lower returns when compared to the stand-alone project.

The case illustrates a scenario wherein (i) a stand-alone project generates healthy returns on investment even while operating at low-efficiency factors, (ii) a sizable market opportunity exists, and enhancing productivity could help meet a part of such market demand, (iii) the increase in productivity is proposed to be achieved at greater than proportional costs, and (iv) the attempts to enhance productivity of one project provides decision makers with options to implement additional projects and to enhance their productivity, thereby spreading the incremental investment in the reservoir across multiple projects.

Case Question

The Via Reservoir provides the option-to-regulate power generation and to enhance productivity. However, if its cost was combined with the investment cost of the stand-alone Mount Coffee Project, the combined project offers lower returns than the stand-alone project. This difference in value creation between two alternative scenarios is the price of the option, and this price has to be made up for by the increased productivity of the intermediate projects.

Teaching Objectives

- Enabling analysts to prepare cash flow estimates for a stand-alone project and a composite project.
- Providing participants with the experience of undertaking a benefit-cost analysis of a real-world situation, involving long construction periods and long-lived project assets.

- Enable analysts to alter input values and assess the sensitivity of key variables.
- For analysts to compare valuations of alternative scenarios and assess the value created by isolated and composite projects.

Case Objectives and Use

The case deals with the analysis of real options that analysts could potentially be faced with. The present case involves public funding for a power plant in a country with barely any generation capacity at all. The stand-alone project proposed generates handsome returns even at modest power tariffs. Yet its production is seasonal and the plant is slated to operate at low-efficiency factors. Policy makers propose the construction of a reservoir to enhance plant usage and output. But the incremental output is generated at a more-than-proportional cost. Hence, the overall returns on the investments generated by the combined project are significantly lower than those earned from the stand-alone project.

In the private sector, managers could undertake value-diminishing projects hoping that a loss-making project would provide the firm with additional market power in the medium to long term. Yet, in this case, the price of electricity is regulated and is likely to be held constant (at least in real terms) for a very long time frame. Worse, this is not a case where complementary product pricing could be employed to recover losses from the sale of consumables or add-ons. Even if the combined project was profitable, it was projected to generate lower returns on investments relative to the stand-alone project. The Via Reservoir is presumed to provide the option-to-launch and to enhance the productivity of intermediate projects. The impairment in value consequent to the inclusion of the reservoir could be considered the price paid for such options. The enhanced productivity from the intermediate projects (besides the enhanced output from Mount Coffee itself) should make up for the reduction in value, and hopefully more.

The instructor could choose to expand the scope of the case to include other aspects that could impact the decision to launch a value-diminishing project owing to the unique characteristic of electricity as an intermediate good, with few substitutes and one that contributes to growth and welfare. In other words, the instructor could hint at investing in a presently loss-making venture with a view to generating social and economic benefits beyond the limits of the projects themselves.

Teaching Plan

The case presents a project that as a stand-alone initiative has received worldwide attention and funding: enhancing efficiency of use of the project assets, not so much. The situation is akin to a puzzle where the first piece is highly profitable by itself but is inadequate to fill the void, the second piece is value-diminishing, and the overall configuration is required to generate greater value to compensate for such impairment in value.

The success of the piecemeal strategy hinges on the promise that the intermediate projects would generate electricity at lower unit costs and the completed puzzle would earn returns on investment comparable to the Mount Coffee stand-alone project, or better. The instructor could extend the discussion to cover various risks involved in such a progression of events, including non-availability of grants from donor nations, higher costs of project debt, inadequate human capacity and resources for expansion and completion etc.

The instructor could guide the students to evaluate breakeven value in a number of small steps as laid out hereinbelow:

Financial Model

1. **Table 1—Revenue**: the revenue model for the Mount Coffee Hydropower Project for 7 years of operations.
2. **Table 1—P&L**: the Earnings After Tax (EAT) for the Mount Coffee Hydropower Project for 7 years of operations.
3. **Table 1—Cash flow**: a statement of cash flows for the Mount Coffee Hydropower Project covering the duration of construction and 7 years of operations.
4. **Table 2—Revenue**: the revenue model for the Mount Coffee Hydropower + Via Reservoir composite project for 7 years of operations.
5. **Table 2—P&L**: the Earnings After Tax (EAT) for the Mount Coffee Hydropower + Via Reservoir project for 7 years of operations.
6. **Table 2—Cash flow**: a statement of cash flows for the Mount Coffee Hydropower + Via Reservoir composite project, covering the duration of construction and 7 years of operations.

The Via Reservoir adds 25 % to Mount Coffee's production, although at a cost higher than that of the Mount Coffee project itself. However, Mount Coffee is the lowest in a planned cascade of five dams/power stations and the reservoir makes it possible to add to the output of each such intermediate project. The proposed intermediate projects—between the Via Reservoir and the Mount Coffee Hydropower Project—need to cumulatively demonstrate value that would justify the reservoir-related investments [over and above the economic returns required to justify their own construction and operation].

Distribution of capex spend: Mount Coffee Hydropower Project (stand-alone)

Year		−4	−3	−2	−1	1
Distribution of spend	%	14.60 %	17.00 %	37.90 %	30.10 %	0.40 %
Distribution of spend	USD	33,580,000	39,100,000	87,170,000	69,230,000	920,000

Project capacity utilization:

Installed capacity = 80 MW × 24 × 365 = 700,800 MWh
Power generated = 432,000 MWh
Plant load factor = 61.64 %

Year		−5	−4	−3	−2	−1	1
Distribution of spend	%	0.52 %	13.58 %	17.23 %	37.86 %	30.29 %	0.52 %
Distribution of spend	USD	2,840,731	73,859,008	93,744,125	205,953,003	164,762,402	2,840,731

Project capacity utilization:

Installed capacity = 80 MW × 24 × 365 = 700,800 MWh
Power generated = 540,000 MWh
Plant load factor = 77.05 %

What Happened Next

Mr. Patrick Sendolo, the Honorable Liberian Minister of Lands, Mines and Energy spoke the significance of the Mount Coffee restoration project at the "ground-breaking" function and said "the restoration of the Mount Coffee hydroelectric plant, which will provide affordable and reliable power to residents and businesses, is an urgent national priority for Liberia that this government has decided to take very seriously." Hon. President Ellen Johnson Sirleaf reiterated her government's commitment to commencing power generation in 2015 and to commissioning the renovated plant by 2016. The Liberian government invited bids for the construction of substations and transmission lines linking the project to the national as well as the West African grid (Hydroworld 2014).

The cash flow chart for the project would need to be drawn up one step at a time:

		1	2	3	4	5	6	7
Power production (%)	% of max	80	85.00	90.00	95.00	100.00	100.00	100.00
Max annual production	kWh	3.46E+08	3.67E+08	3.89E+08	4.10E+08	4.32E+08	4.32E+08	4.32E+08
Aggregate technical and commercial losses	22 %	7.60E+07	8.08E+07	8.55E+07	9.03E+07	9.50E+07	9.50E+07	9.50E+07
Net electricity billed/collected		2.70E+08	2.86E+08	3.03E+08	3.20E+08	3.37E+08	3.37E+08	3.37E+08

(continued)

(continued)

		1	2	3	4	5	6	7
Fixed supply tariffs (till year 2041)	USD	0.148	0.148	0.148	0.148	0.148	0.148	0.148
Revenue from the sale of power		3.99E+07	4.24E+07	4.49E+07	4.74E+07	4.99E+07	4.99E+07	4.99E+07

Composite project: revenue model—Mount Coffee Hydropower Plant + Via Reservoir:

		1	2	3	4	5	6	7
Power production	% of max	80	85.00	90.00	95.00	100.00	100.00	100.00
Max annual production	kWh	4.32E+08	4.59E+08	4.86E+08	5.13E+08	5.40E+08	5.40E+08	5.40E+08
Aggregate technical and commercial losses	22 %	9.50E+07	1.01E+08	1.07E+08	1.13E+08	1.19E+08	1.19E+08	1.19E+08
Net electricity billed/ collected		3.37E+08	3.58E+08	3.79E+08	4.00E+08	4.21E+08	4.21E+08	4.21E+08
Levellized tariffs to end-customers (till 2041)	USD	0.148	0.148	0.148	0.148	0.148	0.148	0.148
Revenue from the sale of power		4.99E+07	5.30E+07	5.61E+07	5.92E+07	6.23E+07	6.23E+07	6.23E+07

1. Stand-alone project: revenue model and profit and loss—Mount Coffee Hydro Power Plant

Year		1	2	3	4	5	6	7
Power production	% of max	80	85.00	90.00	95.00	100.00	100.00	100.00
Max annual production	kWh	3.46E+08	3.67E+08	3.89E+08	4.10E+08	4.32E+08	4.32E+08	4.32E+08
Aggregate technical and commercial losses	22 %	7.60E+07	8.08E+07	8.55E+07	9.03E+07	9.50E+07	9.50E+07	9.50E+07
Net electricity billed/collected		2.70E+08	2.86E+08	3.03E+08	3.20E+08	3.37E+08	3.37E+08	3.37E+08
Levelized tariffs to end-customers (till 2041)	USD	0.148	0.148	0.148	0.148	0.148	0.148	0.148
Revenue from the sale of power		3.99E+07	4.24E+07	4.49E+07	4.74E+07	4.99E+07	4.99E+07	4.99E+07
Operation and maintenance	1.0 %	2,300,000	2,300,000	2,300,000	2,300,000	2,300,000	2,300,000	2,300,000
(Incl. renewals, replacements, insurance, G&A)								
EBITDA		3.76E+07	4.01E+07	4.26E+07	4.51E+07	4.76E+07	4.76E+07	4.76E+07
Interest charge		682,500	647,500	612,500	577,500	542,500	507,500	472,500
Depreciation		4,600,000	4,600,000	4,600,000	4,600,000	4,600,000	4,600,000	4,600,000
Earnings before tax		3.23E+07	3.48E+07	3.74E+07	3.99E+07	4.24E+07	4.25E+07	4.25E+07
Tax charge		5.91E+06	6.38E+06	6.84E+06	7.30E+06	7.76E+06	7.77E+06	7.78E+06
Earnings after tax		2.64E+07	2.85E+07	3.05E+07	3.26E+07	3.47E+07	3.47E+07	3.47E+07

Stand-alone project: loan amortization schedule—Mount Coffee Hydropower Plant (7 of the 20 years shown)

	1	2	3	4	5	6	7
Principal at the beginning of the year	50,000,000	47,500,000	45,000,000	42,500,000	40,000,000	37,500,000	35,000,000
Principal repaid during the year	2,500,000	2,500,000	2,500,000	2,500,000	2,500,000	2,500,000	2,500,000
Principal outstanding at the end of the year	47,500,000	45,000,000	42,500,000	40,000,000	37,500,000	35,000,000	32,500,000
Average principal outstanding during the year	48,750,000	46,250,000	43,750,000	41,250,000	38,750,000	36,250,000	33,750,000
Interest charge (1.4 %)	682,500	647,500	612,500	577,500	542,500	507,500	472,500

Stand-alone project: statement of cash flows—Mount Coffee Hydropower Plant

	−4	−3	−2	−1	1	2	3	4	5	6	7
EAT					2.64E+07	2.85E+07	3.05E+07	3.26E+07	3.47E+07	3.47E+07	3.47E+07
Add back depreciation					4,600,000	4,600,000	4,600,000	4,600,000	4,600,000	4,600,000	4,600,000
Deduct principal repayment					−2,500,000	−2,500,000	−2,500,000	−2,500,000	−2,500,000	−2,500,000	−2,500,000
Cash flow from operations					2.85E+07	3.06E+07	3.26E+07	3.47E+07	3.68E+07	3.68E+07	3.68E+07
Capex outflow (equity)	(10,220,000)	(11,900,000)	(26,530,000)	(21,070,000)	(280,000)						
Net cash flow	−10,220,000	−11,900,000	−26,530,000	−21,070,000	28,220,181	30,565,970	32,631,757	34,697,545	36,763,333	36,791,928	36,820,523
IRR (equity)	32.61 %	(20 years)									

2. Composite project: revenue model and profit and loss—Mount Coffee Hydro Power Plant + Via Reservoir

Year		1	2	3	4	5	6	7
Distribution of spend	%	0.52 %						
Distribution of spend	USD	2,840,731						
Power production	% of max	80 %	85.00 %	90.00 %	95.00 %	100.00 %	100.00 %	100.00 %
Max annual production	kWh	4.32E+08	4.59E+08	4.86E+08	5.13E+08	5.40E+08	5.40E+08	5.40E+08
Aggregate technical and commercial losses	22 %	9.50E+07	1.01E+08	1.07E+08	1.13E+08	1.19E+08	1.19E+08	1.19E+08
Net electricity billed/collected		3.37E+08	3.58E+08	3.79E+08	4.00E+08	4.21E+08	4.21E+08	4.21E+08
Levelized tariffs to end-customers (till 2041)	USD	0.148	0.148	0.148	0.148	0.148	0.148	0.148
Revenue from the sale of power		*4.99E+07*	*5.30E+07*	*5.61E+07*	*5.92E+07*	*6.23E+07*	*6.23E+07*	*6.23E+07*
Operation and Maintenance (incl. renewals, replacements, insurance, G&A)	1.50 %	8,160,000	8,160,000	8,160,000	8,160,000	8,160,000	8,160,000	8,160,000
EBITDA		*4.17E+07*	*4.48E+07*	*4.79E+07*	*5.11E+07*	*5.42E+07*	*5.42E+07*	*5.42E+07*
Interest charge		5,850,000	5,550,000	5,250,000	4,950,000	4,650,000	4,350,000	4,050,000
Depreciation		10,880,000	10,880,000	10,880,000	10,880,000	10,880,000	10,880,000	10,880,000
Earnings before Tax		*2.50E+07*	*2.84E+07*	*3.18E+07*	*3.52E+07*	*3.86E+07*	*3.89E+07*	*3.92E+07*
Tax		4.57E+06	5.20E+06	5.82E+06	6.45E+06	7.07E+06	7.13E+06	7.18E+06
Earnings after Tax		*2.04E+07*	*2.32E+07*	*2.60E+07*	*2.88E+07*	*3.16E+07*	*3.18E+07*	*3.21E+07*

Composite project: loan amortization schedule—Mount Coffee Hydropower Plant + Via Reservoir (7 of the 20 years shown)

	1	2	3	4	5	6	7
Principal at the beginning of the year	200,000,000	190,000,000	180,000,000	170,000,000	160,000,000	150,000,000	140,000,000
Principal repaid during the year	10,000,000	10,000,000	10,000,000	10,000,000	10,000,000	10,000,000	10,000,000
Principal outstanding at the end of the year	190,000,000	180,000,000	170,000,000	160,000,000	150,000,000	140,000,000	130,000,000
Average principal outstanding during the year	195,000,000	185,000,000	175,000,000	165,000,000	155,000,000	145,000,000	135,000,000
Interest charge (3.0 %)	5,850,000	5,550,000	5,250,000	4,950,000	4,650,000	4,350,000	4,050,000

Composite project: loan amortization schedule—Mount Coffee Hydropower Plant + Via Reservoir (7 of the 20 years shown)

	-5	-4	-3	-2	-1	1	2	3	4	5	6	7
EAT						2.04E+07	2.32E+07	2.60E+07	2.88E+07	3.16E+07	3.18E+07	3.21E+07
Add back depreciation						10,880,000	10,880,000	10,880,000	10,880,000	10,880,000	10,880,000	10,880,000
Deduct principal repayment						−10,000,000	−10,000,000	−10,000,000	−10,000,000	−10,000,000	−10,000,000	−10,000,000
Cash flow from operations						2.13E+07	2.41E+07	2.69E+07	2.97E+07	3.25E+07	3.27E+07	3.29E+07
Capex outflow (equity)	(1,221,932)	(31,770,235)	(40,323,760)	(88,590,078)	(70,872,063)	(1,221,932)						
Net cash flow	−1,221,932	−31,770,234	−40,323,759	−88,590,078	−70,872,062	20,066,793	24,080,316	26,871,907	29,663,498	32,455,089	32,700,189	32,945,289
IRR (equity)	9.916 %	(20 years)										

References

Alfaro J, Miller S (2014) Satisfying the rural residential demand in Liberia with decentralized renewable energy systems. Renew Sustain Energy Rev 30:903–911 (Elsevier Limited)

Appleyard D (2014) Hydropower 2014 outlook: hydro industry to expand its global reach, *Renewable Energy World*, January/February 2014. http://www.renewableenergyworld.com/rea/news/article/2014/01/hydropower-2014-outlook-hydro-industry-to-expand-its-global-reach. Accessed 31 Jan 2014

Gaillard RM (2014) Presentation of results evaluation of EU energy funding in Liberia and recommendations, 24 July 2012, MWH. Accessed 28 Jan 2014

Hydroworld (2014) Liberia breaks ground on mount coffee hydropower plant rehabilitation. www.hydroworld.com, 30 Jan 2014. http://www.hydroworld.com/articles/2014/01/liberia-breaks-ground-on-mount-coffee-hydropower-plant-rehabilitation.html?cmpid=EnlHydroFebruary42014. Accessed 3 Feb 2014

The Liberian Observer (2014) President Sirleaf breaks ground Saturday, Kicking off mount coffee hydro project, 24 January 2014. http://www.liberianobserver.com/news/president-sirleaf-breaks-ground-saturday-kicking-mount-coffee-hydro-project. Accessed 27 Jan 2014

Wesseh PK Jr., Zoumara B (2012) Causal independence between energy consumption and economic growth in Liberia: evidence from a non-parametric bootstrapped causality test. Energy Policy 50:518–527 (Elsevier Limited)

Chapter 11
Nepal, Ridi Hydro: Common Equity—Uncommon Enthusiasm

Valuing A Public Offer of Equity Shares

The risks are less and the returns are larger in hydropower companies compared to other sectors.
—Rabindra Bhattarai, Nepalese Stock Market Analyst

Most of these investors are from the secondary market and they invest their money with the expectations of good return in short span of time.
—Anjan Raj Poudel, Former President Stock Brokers' Association of Nepal

The issuance of primary shares will help to change the perception that Nepal's hydropower companies are not transparent.
—Kuber Mani Nepal, Director Ridi Hydro Power Development Company

Background

The offer of equity shares to the public by Nepalese energy generation major, Ridi Hydro Power Development Company (ridihydro.com.np), was oversubscribed by a factor of 92. Applicants had paid in over Nepalese rupee (NPR) 11 billion for 1.17 million equity shares of face value NPR 100 ($\sim$ USD 0.99). The public issue opened on Sunday (the first business day of the week in Nepal), February 16, 2014, and closed on Thursday, February 20, 2014, four days later. Credit rating agency ICRA Nepal (icranepal.com) had assigned "Grade 3" ("average fundamentals") to the company's offering on a 5-point scale, with Grade 1 representing strong fundamentals. Individual applicants could subscribe to a minimum of 50 and a maximum of 10,000 (5,000 for project-affected local residents) equity shares (Fig. 11.1).

The company had, in October 2013, issued 0.3 million equity shares to residents impacted by the 2.40-MW Ridi Khola (=river) small hydro power (SHP) project. 58,500 of the shares were reserved for selected mutual funds and 23,400 for the company's employees (ShareSansar 2014). Locals subscribing to the issue were required by law to hold the allotted securities for a lock-in period of 3 years.

© Springer India 2015
S. Sunderasan, *Cleaner-Energy Investments*,
DOI 10.1007/978-81-322-2062-6_11

Fig. 11.1 2.40-MW Ridi Khola small hydro power project: *picture credit*—sharesansar.com

Project Description

The company's project assets are listed in the table below and the projects had been operating at an average annual plant load factor of 65 % (Table 11.1).

The Nepal Electricity Authority (NEA: www.nea.org.np), the sector regulator and transmission and distribution utility, had opened up generation to private sector participation. The government had developed high-voltage transmission corridors to evacuate power from the various hydropower projects. The NEA enjoyed a healthy track record of payments, often making payments to generators within 90 days of drawal, and hence, offtake and payment risks were minimal. Power from the Iwa Khola project, due for commercial launch in July 2018, was to be evacuated first by a 10 km/132 kV transmission line to the *Kabeli* substation and then conveyed to the *Damak* substation on a 90-km line: each of these lines and the *Kabeli* substation were to be built by the NEA and delays in execution could affect project financial performance.

Table 11.1 List of projects managed by Ridi Hydro Power Development Company Limited inputs sourced from: http://sharesansar.com/viewnews.php?id=18937

Project name	Capacity (MW)	Location (District)	Status	Capital cost (NPR million)	Contracted energy (GWh/ year)
Ridi Khola	2.40	Gulmi and Palpa	In operation (Oct 2009)	400.00	15.730
Piluwa Khola	3.00	Shankuwasava	In operation (Aug 2003)	326.40	19.560
Iwa Khola	9.90	Panchthar and Taplejung	Under construction (expected July 2018)	1,494.80	56.625
Rairang (acquired and managed by Ridi Hydro)	0.50	Dhading	In operation (2004)/renovated 2012		2.320

Investors were not exposed to project completion risks or cash flow risks as the company had three operating projects aggregating about 5.90 MW and was on track to commissioning the 9.90-MW Iwa Khola project by November 2017. The company had reported profits of NPR 21.2 million ($\sim$ USD 0.22 million) for the fiscal year ended July 15, 2013.

Companies desirous of tapping into the capital markets to mobilize equity from the public in Nepal were mandated, by law, to issue shares at face value. Investors, therefore, hoped to make sizable arbitrage profits when trading in the shares commenced on the exchange (Ghimire 2014).

Ridi Hydro was the sixth hydropower company to list on the Nepal Stock Exchange (www.nepalstock.com). Sanima Mai (www.maihydro.com) that had made its stock market debut barely 2 weeks prior in mid-January 2014 was trading at NPR 500[1] at the time of the Ridi share offering.

Post-issue of equity shares to the public, the shareholding of the company was divided between promoters and non-promoters in the ratio of 51:49 %. Arun Valley Hydropower Limited (another listed and publicly traded entity) was the single largest shareholder (16.5 %), followed by other promoter individuals and their family investment companies (34.5 %), project-affected area residents (10 %), and members of the public (39 %). The company had proposed to use the proceeds from the share sale to repay Ridi Khola project debt and to invest in the construction of the Iwa Khola project. The promoters brought in considerable project execution and consulting experience to be able to manage timely execution of the Iwa project.

In addition to the anticipated windfall profits on listing, investors also took a liking to hydropower stocks because the returns on offer were not cyclical and were relatively unaffected by short-term interest rate fluctuations, when compared to the banking stocks that dominated the Nepal Stock Exchange (NMB Capital Limited 2014). Dry season (NPR 8.4/kWh) and wet season (NPR 4.8/kWh) tariffs were predetermined with a 3 % escalation for five years, and returns on equity were generally stable and predictable. Investment bankers considered hydropower to be a low-β stock (low on volatility) and Ridi Hydro specifically was seen as a company with ample opportunity for reinvestment and growth. Hence, when stock prices stabilized after the listing frenzy, common equity shares in hydropower companies performed almost like fixed-income securities, offering stable, and often healthy, dividends to the shareholders.

Hydropower companies also enjoyed a ten-year tax holiday from commencement of commercial operations, followed by a discounted tax of 10 % for five years and 20 % for years 15–20. The operating costs were a modest 3 % of revenues and hydropower projects were, by and large, expected to earn 97 % in gross profit margins. The company's product (electricity) was not faced with threats of substitutes, and diesel, solar PV, and even wind energy were higher marginal cost competitors.

---

[1] Ridi to Be Sixth Listed Hydro Firm at Nepse, 4 February 2014, http://www.newsuk24.com/news/ridi-to-be-sixth-listed-hydro-firm-at-nepse, last accessed 3 April, 2014.

Even as marginal costs of generation were among the lowest, on the down side, hydrokinetic generation offered near-constant returns to scale. Abnormal returns, relative to installed capacity and expected efficiency factors, were highly unlikely. The company had proposed to pay down its debt after the IPO, and by July 2014, its outstanding debt was to be at 66 % of capital, further reduced to 51 % by July 2016. In the past, Ridi Hydro had been prompt in servicing its debt to the lender consortium (ICRA 2014). Even as the forecast debt-to-equity ratios were quite low, and as the company was unlikely to be adversely affected by moderate costs of debt, higher interest rates on project debt (of the order of 15 % or higher) could potentially challenge the company's liquidity.

Means of Financing

The capital structure proposed for the 9.90-MW Iwa Khola project was 75 % in long-term debt and 25 % in equity. The February 2014 IPO was to contribute a portion of this equity. Ridi Hydro hoped to contribute additional equity to the project from internal accruals from existing/operating plants. The company had also proposed to float shares, possibly of a special purpose vehicle constituted at a later date, to hold the Iwa project assets. With a view to erring on the side of caution, for the purpose of the present valuation, the cash flows projected to accrue from the *Iwa* project could be ignored. At a later date, the Ridi Hydro could receive dividends from the *Iwa* special purpose vehicle, in proportion to its shareholding in the latter, but the said stream of cash is ignored to try and arrive at a conservative valuation.

Project Data

Key Statistics
All figures in NPR million (except per share data)

Year (15 July–14 July)	10/11	11/12	12/13	13/14[a]	14/15[a]	15/16[a]
Interest expenses	33.114	30.921	26.664	25.415	20.820	18.427
Net profit	10.721	8.472	21.203	35.684	39.837	35.767
Investments		31.500	31.555	121.500	131.500	131.500
Long-term debt	254.154	197.360	225.260	198.296	175.496	152.693
Equity capital	152.500	152.500	153.000	300.000	300.000	300.000
Reserves	2.076	0.941	2.866	3.855	7.838	
Debt–equity	167 %	129 %	147 %	66 %	58 %	51 %
Earnings per share	7.00	5.50	13.90	11.90	13.30	11.90
Dividend per share	5.20	6.30	12.60			

[a] Projected

Teaching Note

Case Synopsis

Companies seeking to tap into the capital markets by issuing common equity shares to the public in Nepal were mandated by law to issue such shares at face value, typically of Nepalese rupee (NPR) 100. Such pre-money valuation ignored the windfall gains likely on listing, for, when trading in the shares commenced, they were typically priced at a substantial premium over the issue price based on post-money valuation. Hydropower stocks in Nepal offered a welcome diversification for investors in a market dominated by banking and finance companies. Marginal costs of hydrokinetic generation were negligible: the income streams were considered stable and reasonably predictable. The market for the product (electricity) was, for most part, without substitute, almost unlimited, both at home in Nepal as well as for export to neighbors, viz. India and Bangladesh.

Ridi Hydro Power Development Company Limited operated 5.90 MW of hydroelectric generation capacity, spread across 3 projects, and was well into the construction of the 9.90-MW *Iwa Khola* project that was expected to commence commercial operations by July 2018. On average, hydropower plants in Nepal were expected to operate at an efficiency of about 65 %. The first quarter of the calendar year (January–March) was the dry season and generators enjoyed a higher tariff of NPR 8.4 per kWh, while for the remaining 9 months, the tariff was fixed at NPR 4.8 per kWh. The Nepal Electricity Authority was both the sector regulator and the transmission and distribution utility and had enjoyed a healthy record of payments to generators: payments for power drawn were made within about 3 months.

The shares were offered at face value of NPR 100 amounting to 39 % of the common equity shares issued. The initial public offering (IPO) was subscribed over by a factor of 92. Observers had believed that this enthusiasm for the issue was primarily driven by the opportunity to derive windfall gains on listing. Recently listed hydropower companies had been trading at 4–5 times the face value. Others believed that hydropower investments represented a long-term risk-free option as cash flows to operators, and hence, dividends to shareholders were generally stable and predictable.

Case Question

What would be the estimated firm value and hence the arbitrage-free pricing of the public offer of shares for Ridi Hydro.

Teaching Objectives

- Enabling students to study analyst reports of an initial public offer and to prepare their own versions of the cash flow estimates.
- Enable students to study the risk factors and to alter input values and assess the sensitivity of key variables.
- For students to compare valuations of alternative scenarios and assess the arbitrage value generated by the legal mandate.

Case Objectives and Use

The case deals with the valuation of a firm making an initial public offering of common equity shares. The complexity is limited to the extent that the efficiency factors, electricity output from the power plants, the tariff patterns, and the revenue streams were expected to be stable and predictable. Upon repayment of project debt, the company was slated to be left with larger free cash reserves that could be paid out as dividends (or reinvested into subsequent projects). To a limited extent, the equity share, in fact, was valued along the lines of a fixed-income security.

The mandate to issue shares at face value (pre-money) rather than on a post-money basis created an arbitrage opportunity and hence attracted potential investors in large numbers. The case intends to highlight this situation. The shares were actually issued by a draw of lots, and winners received windfall gains on listing, akin to other investments whose returns are determined by chance, rather than by insight or skill. In the long run, however, hydropower companies were expected to yield stable and predictable dividend flows and hence to attract long-term investors.

The instructor could steer discussions to view the arbitrage opportunity as an aberration and to help compute the value of the firm from free cash flows. On the other hand, the instructor could also discuss the nature of the hydropower resource as a public asset and bring out the moral justification of issuing shares at face value to the residents affected by the project, the company staff, and eventually to the public. In that sense, the company would be presumed to hold the asset in the spirit of a public trust, and the benefits from the project would be seen as accruing to society at large (or at the very least to those receiving an allotment of shares by the draw of lots).

Teaching Plan

By its very nature, hydrokinetic power generation offers constant returns to scale, and there is generally little that managers could do to dramatically enhance efficiency of operations or electricity output. The inputs to the valuation are derived from analyst reports which, in turn, are based on a set of actual accounting data and projections.

The instructor should help draw up a projected loan amortization schedule, and a free cash flow model to help arrive at the value of the firm as a whole. Based on this valuation of the firm, the value of each share could be ascertained.

The instructor could guide the students to evaluate the price band for the post-money value of the shares, subject to assumptions regarding the future, as laid out hereinbelow:

Financial Model

1. **Table 1—Project Cost**: average cost per kW of project construction (to help estimate depreciation figures).
2. **Table 2**—Listing of relevant inputs to the financial model.
3. **Table 3—Loan Amortization Schedule**: a statement showing the actual and projected repayment of project debt and the interest charge thereon (feeding into the project's P&L account).
4. **Tables 4, 5 and 6—Cash flow estimation and firm valuation**: statements of cash flows for various valuation scenarios based on available and projected input data.

Table 1: Average cost of project construction

Name of the project	Cost (NPR million)	Capacity (kW)
Ridi Khola	400.00	2,400
Piluwa Khola	326.40	3,000
Iwa Khola	1,494.80	9,900
Total	2,221.20	15,300
Average (NPR million/kW)	*0.145176471*	

Table 2: Inputs to the cash flow model

Power generation capacity (kW)	5,900
Average PLF (%)	65 %
Annual power generation (kWh)	33,594,600
Dry weather tariff (3 months) (NPR)	8.4
Wet weather tariff (9 months) (NPR)	4.8
Project life (year)	25
Project cost (5.90 MW) (NPR million)	856.5411765
Depreciation (SLM/25 years) (NPR million)	34.26164706
Expected return on equity (%)	15.00

Table 3: Projected loan amortization schedule

	1	2	3	4	5	6	7	10
	2012–2013[a]	2013–2014[b]	2014–2015[b]	2015–2016[b]	2016–2017[c]	2017–2018[c]		2021–2022[c]
Debt outstanding at year end	225,260,000	198,296,000	175,496,000	152,693,000	129,890,000	107,087,000	–	15,875,000
Principal repaid during the year	–27,900,000	269,64000	228,00000	228,03000	228,03000	228,03000	–	15,875,000
Average debt for the year	211,310,000	211,778,000	186,89●,000	164,094,500	141,291,500	118,488,500	–	27,276,500
Interest charge	26,664,000	25,415,000	20,820,100	18,427,000	15,866,336	13,305,672	–	3,063,016
Interest cost (%)	12.62	12.00	11.14	11.23	11.23	11.23	–	11.23

[a] Actual
[b] Analyst's projections
[c] Author's projections

Table 4: Cash flow model/scenario 1 (terminal value at 30x of Y5 free cash flow)

	2013–2014	2014–2015	2015–2016	2016–2017	2017–2018	Terminal value
Total revenue	191,489,220.00	191,489,220	191,489,220	191,489,220	191,489,220	30x Y5 cash flow
Interest expenses	25,415,000.00	20,820,000.00	18,427,000.00	18,427,000.00	18,427,000.00	
Net profit	*35,684,000.00[a]*	*39,837,000.00[a]*	*35,767,000.00[a]*	*37,096,000.00[b]*	*37,566,666.67[b]*	
Add back depreciation[b]	34,261,647.06	34,261,647.06	34,261,647.06	34,261,647.06	34,261,647.06	
Deduct principal repayment	26,964,000.00[a]	22,800,000.00[a]	22,803,000.00[a]	22,803,000.00[b]	22,803,000.00[b]	
Net cash flow	42,981,647.06	51,298,647.06	47,225,647.06	48,554,647.06	49,025,313.73	1,470,759,412
Firm value (as of January 2014)	852,660,967.40					
Shares issued to the public	1,170,000	39 %				
Total shares issued	3,000,000	100 %				
Price per share	**284.22**					

[a] Analyst's projections
[b] Author's projections; 2013–2014 free cash flow discounted for 6 months, 2014–2015

Table 5: Cash flow model/scenario 2 (terminal value at 20x of Y5 free cash flow)

	2013–2014	2014–2015	2015–2016	2016–2017	2017–2018	Terminal value
Total revenue	191,489,220.00	191,489,220	191,489,220	191,489,220	191,489,220	20x Y5 cash flow
Interest expenses	25,415,000.00	20,820,000.00	18,427,000.00	18,427,000.00	18,427,000.00	
Net profit	*35,684,000.00*[a]	*39,837,000.00*[a]	*35,767,000.00*[a]	*37,096,000.00*[b]	*37,566,666.67*[b]	
Add back depreciation[b]	34,261,647.06	34,261,647.06	34,261,647.06	34,261,647.06	34,261,647.06	
Deduct principal repayment	26,964,000.00[a]	22,800,000.00[a]	22,803,000.00[b]	22,803,000.00[b]	22,803,000.00[b]	
Net cash flow	42,981,647.06	51,298,647.06	47,225,647.06	48,554,647.06	49,025,313.73	980,506,275
Firm value (as of January 2014)	625,369,956.84					
Shares issued to the public	1,170,000	39 %				
Total shares issued	3,000,000	100 %				
Price per share	**208.46**					

[a] Analyst's projections
[b] Author's projections; 2013–2014 free cash flow discounted for 6 months, 2014–2015

Table 6: Cash flow model/scenario 3 (terminal value at 10x of Y5 free cash flow)

	2013–2014	2014–2015	2015–2016	2016–2017	2017–2018	Terminal value
Total revenue	191,489,220.00	191,489,220	191,489,220	191,489,220	191,489,220	10x Y5 cash flow
Interest expenses	25,415,000.00	20,820,000.00	18,427,000.00	18,427,000.00	18,427,000.00	
Net profit	35,684,000.00[a]	39,837,000.00[a]	35,767,000.00[a]	37,096,000.00[b]	37,566,666.67[b]	
Add back depreciation[b]	34,261,647.06	34,261,647.06	34,261,647.06	34,261,647.06	34,261,647.06	
Deduct principal repayment	26,964,000.00[a]	22,800,000.00[a]	22,803,000.00[a]	22,803,000.00[b]	22,803,000.00[b]	
Net cash flow	42,981,647.06	51,298,647.06	47,225,647.06	48,554,647.06	49,025,313.73	490,253,137
Firm value (as of January 2014)	398,078,946.28					
Shares issued to the public	1,170,000	39 %				
Total shares issued	3,000,000	100 %				
Price per share	**132.69**					

[a] Analyst's projections
[b] Author's projections; 2013–2014 free cash flow discounted for 6 months, 2014–2015

References

Ghimire S (2014) Ridi hydropower's IPO oversubscribed by 92 times, Republica 23 Feb 2014. http://myrepublica.com/portal/index.php?action=news_details&news_id=70044. Accessed 3 Apr 2014

ICRA (2014) ICRA Nepal assigns IPO grade 3 to the proposed IPO issue of Ridi Hydropower Development Company Limited ICRA Nepal, Dec 2013. http://www.icranepal.com/RecentReleases/Ridi%20Hydropower%20Rationale%20Final.pdf. Accessed 3 Apr 2014

NMB Capital Limited (2014) The analyst: initial public offering | NEPSE 804, 17 Feb 2014. http://www.nmbcapital.com.np/uploads/NMBCL_Ridi_IPO_20140217.pdf. Accessed 3 Apr 2014

ShareSansar (2014) Ridi hydropower Issues IPO of 11.70 Lakh Units from Today, ShareSansar, 16 Feb 2014. http://sharesansar.com/viewnews.php?id=18937&cat=news. Accessed 3 Apr 2014

Chapter 12
Wyke Farms: Sustaining Personality and Profitability

Monetizing Greener Branding

Dairy farms have not typically been set up with energy efficiency in mind and Wyke Farms is unique in its dynamic approach. We are delighted it is meeting its objectives earlier than expected.
—Jeff Ingvaldson, Managing Director—Brilliant Harvest

We aim to operate our business in a way that has minimal impact on the Somerset environment and to create a truly symbiotic relationship with the countryside that provides our food, our income and our home.
—Rich Clothier, Managing Director and 3rd generation family member at Wyke Farms

It's great to see a family farming business like Wyke Farms investing in such an innovative green energy scheme. Making use of their waste products in order to become 100 % self-sufficient in green energy will make them more efficient and help the environment.
—David Cameron, Honorable Prime Minister of Great Britain and Northern Ireland

Background

Farmhouse cheddar was no average cheese. The company that produced it believed that its product was "extra special." The UK/Somerset, based Clothier family continued to produce the cheddar range at Wyke Farms using Ivy Clothier's original recipe, first created in 1861. The cheese range was branded "Just Delicious Extra Mature," "Simply Gorgeous Vintage," "Rich & Creamy Mature," and "Super Light," each sold at GBP 12.0 for 900 g including shipping, on the company's online store, www.wykeshop.com (Fig. 12.1).

The *extra mature* cheddar peaked over 12 months and was one of the company's best-selling variants, frequently referred to as "The Queen of Cheddars." The *vintage* cheddar was matured over a 15-month period by master cheese makers and had been awarded the first place (gold) among over 2,000 competing cheeses at the

© Springer India 2015
S. Sunderasan, *Cleaner-Energy Investments*,
DOI 10.1007/978-81-322-2062-6_12

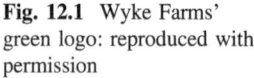

Fig. 12.1 Wyke Farms'
green logo: reproduced with
permission

100% Green

2007 International Food Exhibition in London. The *Rich & Creamy* variant was accorded the "Best Mature Cheddar" in the world in 2004. In keeping with the times and with a view to offering a low-cholesterol cheddar variant for growing kids and health-conscious adults, Wyke designed the *super light* with a mere 1.3 % saturated fat, with vitamin E, protein, sodium, and with calcium levels comparable to mainstream cheeses.

Consumption of organic dairy products in the UK grew at 13.7 %, compounded annually, in inflation-adjusted dollar terms, for the period 2005–2010 and was projected to grow at 1 % annually during 2010–2015.[1] The cheese market was intensely competitive and discounting had been extensive, and retailers also sold imported, own-labeled and valued product. To put things in perspective, the online portal of supermarket chain Tesco (www.tesco.com) retailed[2] rival *mature* cheeses at about GBP 6.40 and *vintage* cheeses at about GBP 10.10 for 900 g. Waitrose (www.waitrose.com) had offered *mature, extra mature,* and *vintage* cheddar between GBP 9.50 and 10.50 per kilogram.[3] The Wyke cheddar variants themselves were retailed at prices ranging between GBP 10.0 and 13.34 per kilogram. Asda (www.asda.com) retailed *medium white, mature,* and *extra mature* cheddar at GBP 6.09 per kilogram.[4] Despite the premium pricing, as of September 2013, the company had claimed that Wyke was the fastest growing cheddar brand in the UK, with a 10 % year-on-year growth in recent years.

---

[1] "Organic Packaged Food in the United Kingdom," Market Indicator Report, Agriculture and Agri-food Canada, November 2011, p 7 and 8 of 22, http://www.ats-sea.agr.gc.ca/eur/6058-eng. htm, last accessed March 24, 2014.

[2] http://www.tesco.com/groceries/product/search/default.aspx?notepad=cheese&N=4294796065 +0&Nao=0, last accessed March 24, 2014.

[3] http://www.waitrose.com/webapp/wcs/stores/servlet/JotterResultsView?jotteritem=cheddar, last accessed March 24, 2014.

[4] http://groceries.asda.com/asda-webstore/landing/home.shtml?cmpid=ahc-_-ghs-sna7-_-asdacom-dsk-_-hp&#/aisle/910000975464, last accessed March 24, 2014.

Wyke Farms was the UK's largest independent milk processor and cheese producer with an annual sales volume of over 14,000 t. But the company was known for more than just its size. Wyke's barn at its Champflower Dairy held its largest herd of cows. It was also fitted with 117 numbers of solar PV modules generating over 39 kW/28,000 kWh of electricity a year. The 49.92 kW solar PV array, the company's first solar PV installation comprised 208 numbers of 240 W panels and 2 inverters, generating 46,200 kWh of energy a year. Surplus energy was used to heat water on the farm. The solar PV installations were expected to pay for themselves in avoided electricity costs alone in a mere 5–6 years.[5]

While these attempts in themselves set Wyke Farms apart from the competition, what made it truly unique was that it became UK's first national cheddar brand to be 100 % energy self-sufficient, with the launch of its biogas plant in September 2013.

Project Description

The GBP 5.0 million biogas plant, consisting of three numbers of 4,600 m^3 on-site anaerobic digester vessels and two gas engines, took 5 years to plan, build and commission, and was designed to run on 75,000 t of cow manure and other waste material from the farm. In addition to the energy generated, it reduced labor invested into managing waste (Stones 2014). Each day, the digester tanks were fed with 150 t of cow and pig manure, plant stalks, and other biodegradable waste feedstock. The whey permeate from cheese making was used as feedstock. The whey was high in sugar and was fed slowly to avoid surges in gas production. Some oxygen was also injected into the digester header tanks to encourage the growth of bacteria.

The project was expected to generate some 250 m^3 of biogas per hour. This gas was a mixture of methane, carbon-dioxide, water vapor (moisture), and sulfur-dioxide. It was dehydrated and scrubbed for sulfur-dioxide (enriched) and then combusted in 670-hp/499 kWe gas engines. Enriching the gas prevented production of sulfuric acid and the consequent corrosion (Appleyard 2013). A heat-recovery system was installed to make the most of the waste heat from the engines and recycled the heat to the digester tanks. Plastic tubes circulated the heated water from the heat exchanger, maintaining a 40 °C temperature, to promote biological activity. Feed water and other source materials were also pre-warmed. In the manner of completing the materials cycle, the pasteurized natural fertilizer resulting from the digestion process was plowed back into the pasture lands, displacing chemically produced nitrogen fertilizer. As a backup option, gas burners were to be employed to flare the gas in the event of engine failures.

---

[5] "Brilliant Harvest Installs Solar Panel System for Wyke Farms as cheesemaker goes 100 % Green", http://www.sourcewire.com/news/80604/brilliant-harvest-installs-solar-panel-system-for-wyke-farms-as, last accessed March 24, 2014.

Project Financing

Under a 20-year power purchase agreement (PPA), the project received GBP 0.1402/kWh (USD 0.087/kWh) under the UK's renewable energy feed-in tariff scheme, for energy produced and exported to the utility grid. The design plant load factor of 96 % and above meant that annual support of about GBP 1.20 million (USD 2 million) accrued to the company, representing a simple payback of under 5 years.

In all, Wyke Farms had put together a GBP 10 million (~USD 16 million) green energy and environmental master plan including proposals for water recovery from dairy operations, upgrading the biogas to make it suitable for transmission through the national gas grid, additional waste heat-recovery systems, and opening a *Green Visitor Center*. An electric vehicle charged through the on-site solar PV installations was employed for local deliveries.

Wyke Farms was mindful of the possibility that eventually the competition would catch-up to replicate their efforts at using farm wastes and manure for power generation, diluting their first-move brand differentiation. The company possibly had a 3- to 5-year window to recoup the branding-related investment.

Adding Color to the Brand

- The cheese making process itself was standardized, especially cheddar, gouda, and other mass-produced cheeses, and in the absence of opportunities for value addition, competition had driven margins down dramatically.
- Some regulators were mulling taxing high-fat cheese alongside unhealthy consumer products such as alcohol and tobacco, putting conventional cheese at a disadvantage (Armenian Development Agency 2012).
- Consumers were found willing to pay more for regionally produced speciality cheeses, especially those protected with registered geographic identifications or designations of origin.
- Consumers were willing to pay for quality and taste, when combined with a strong marketing appeal. The Wyke brand was known for its history, and for the traditional cheese making processes that were handed down over four generations.
- However, Wyke was too large operation to serve a niche speciality cheese audience and retailed through organized multi-product retail chains.
- Many of the dominant players across the European Union had focused on new product development, line extensions, and innovations in packaging: The new products included cheese snacks that children could take to school.
- Smaller producers developed the "embedded-cheese" segment for customers serving pizzas, snacks, and sauces. This involved developing dedicated ingredient solutions with different melting points, different shapes, and cheese powders. The intellectual property associated with some of these new processes and products was protected and served as a competitive edge.

- Intense competition had also forced cheese makers to innovate on convenience of use, to improve portability, and to try and increase the frequency of cheese consumption.
- Traditional business models and strategies were constantly challenged and survival and growth meant more than having just production at scale and at a competitive price.
- Wyke Farms adopted a zero-net energy strategy to enhance efficiency and to differentiate its brand recall, creating a unique selling proposition and a contemporary story to strengthen its legacy. Richard Clothier, Managing Director was quoted saying, "I own this business and I believe in our brand which stands for more than simply quality, but for sustainable business in its fullest sense A brand is more than simply a promotional label. Our brand has substance. It has provenance, history and real values..." (Gray 2013).

	March 2013	March 2012	March 2011	March 2010	March 2009
Turnover	59,820,000	59,260,000	59,075,000	73,170,000	74,579,000
Pre-tax profit	2429000	2181000	1243000	−879000	1,124,000
Pre-tax profit margin (%)	*4.06*	*3.68*	*2.10*	*−1.20*	*1.51*
Post-tax profit	1,817,000	1,670,000	931,000	−573,000	1,017,000
Balance sheet					
Net assets	13,950,000	12,442,000	11,056,000	10,281,000	11,387,000
Total assets	43,868,000	38,017,000	33,722,000	38,167,000	42,427,000
Total liabilities	29,918,000	25,575,000	22,666,000	27,886,000	31,040,000
Tangible assets	175,333,000	13,540,000	12,432,000	12,667,000	12,897,000
Intangible assets	*10,000*	*10,000*	*10,000*	*115,000*	*140,000*
Fixed assets	175,343,000	13,550,000	12,442,000	12,782,000	13,037,000
Current assets	26,325,000	24,467,000	21,280,000	25,385,000	29,390,000
Stock	18,658,000	16,686,000	13,643,000	14,890,000	18,895,000
Other debtors	785,000	631,000	359,000	835,000	713,000
Current liabilities	11,222,000	22,378,000	12,628,000	15,919,000	27,681,000
Trade creditors	7,897,000	6,630,000	5,356,000	6,424,000	9,684,000

Note All figures in GBP

Teaching Note

Case Synopsis

The Clothier family had produced cheese for four generations employing traditional practices and, over time, their company, Wyke Farms had evolved into the UK's largest family owned, independent milk processor and cheese producer. Despite the legacy and the brand recall, the company found itself competing in a crowded and commoditized market place. Wyke Farms decided to invest GBP 10 million in renewable energy systems and to produce all the electricity it used from farm manure and spent wash from cheese making. Toward this end, the company installed biogas digesters, gas engines, an electric passenger car, and solar photovoltaic systems. In the manner of closing the materials loop, the digested manure emanating from the biogas reactors was spread over the pasture lands as a natural fertilizer. In all, the company's efforts were directed at 100 % self-reliance for energy generation and at minimizing impact on the natural environment. The investments were to pay for themselves in about 6 years from avoided electricity and fuel costs.

The secondary payoff from the investment was also to add character to a brand which was already rich with legacy. The publicity surrounding the green-energy-related investments and the guided tours of the visitors' center, including one by Mr. David Cameroon, the Prime Minister of Great Britain and Northern Ireland, created a favorable image and helped the company with a new narrative. The initiative was believed to have strengthened the brand identity and was expected to contribute to superior brand equity.

Yet this advantage would have been short lived and the company would have to move fast to monetize the equity. In fact, Wyke itself encouraged other farms to adopt similar environment-conscious production strategies. The case discusses a green-branding scenario and the possibility of monetizing such branding and the relatively brief time window available to first movers to recover investments made into such green branding. In the medium-term such practices could emerge as industry norms and individual firms would need to look for newer techniques to differentiate their brands.

Case Question

To estimate the brand equity derived by Wyke Farms from the green-energy-related investments and the longevity of such differentiation.

Teaching Objectives

- Enabling students to appreciate the non-financial returns from energy-related investments.
- Providing students with the experience of undertaking a benefit-cost analysis of investments into intangible assets.
- Enabling students to assess the risks involved in such investments and to design appropriate mitigation measures.
- For students to compare valuations employing alternative methodologies and to assess the longevity of the value created by the cleaner energy investments.

Case Objectives and Use

The case deals with the analysis of clean energy investments that earn returns across several facets of business operation. The biogas digesters eliminated the need for labor to dispose off the manure on the farm. The avoided cost of electricity displaced by the digesters would pay for the project in 5–6 years. But most significantly, the green-energy-related investments provided the company with a new narrative to attach to the 150 year legacy of cheese making.

In the private sector, managers invest in projects that add character to a brand and help differentiate the product, hoping that such projects would provide the firm with additional market power in the medium to long term. When the physical identity of a brand is clearly established in the minds of the consumers, the brand could assume a larger identity and carry a social and environmental message in the manner of contemporary "cause-related branding" initiatives. Yet, in this case, the time window available for the company to be able to monetize the greener branding was relatively short. The cheddar market was generally commoditized and margins were likely to remain slim. The company was too large to serve the premium segment with a niche offering at premium pricing. The product was retailed through groceries and large super market chains.

The market mix of a product is generally defined by its "four Ps": product, pricing, promotion, and positioning. In this case, the cheddar product was fairly commoditized and Wyke Farms had already introduced the low-fat version to appeal to the calorie conscious segment of the market. The pricing was marginally higher than comparable product retailed by supermarkets and their online operations. It was positioned as a "British" indigenous brand in the UK and was promoted through most traditional and modern media including through social networks.

The instructor could highlight the fact that the company had chosen to highlight its process (traditional and unchanged for over a century) and had expanded its scope to environment-friendly production. It brought a new dimension to the brand and highlighted this new character at every available opportunity. It demonstrated the company's commitment to minimizing the impact of its operations on the environment. The environmental benefits notwithstanding, the green energy projects would have paid for themselves through avoided manure management and through electricity displaced. Yet, "100 % self-sufficiency in energy" was a powerful branding initiative.

Teaching Plan

The case covers a unique project in that the investment is made into clean energy infrastructure but its returns are earned from product differentiation and superior brand recall. The instructor could apply the market approach and benchmark the brand's value against transactions involving similar product sold under other brands. The alternative would be to use the cost approach: The costs involved in creating a brand of identical recall. In this case, however, since the brand already enjoyed a legacy and a strong following, the counter-factual scenario would need to be constructed and the incremental value derived from the clean energy investments would need to be computed. Finally, the instructor could employ the income approach that estimates the present value of the future income attributable to the branding. The incremental earnings under these circumstances could accrue from premium pricing and healthier margins.

The analysis could then be expanded to cover the entire business at Wyke Farms and the value accruing to the shareholders on account of the invigorated branding. Consumers believe that their choice of brand, especially relating to products of conspicuous consumption, makes a statement about their own personalities and convictions. Preferring certain brands over others confers emotional benefits to the consumer. All other things being equal, branding enhances the reality behind the product: In this case, the production process. Yet, if the company were to price the product far ahead of the competition, it could lead to a breakdown in consumer confidence. Hence, the company was required to balance between premium pricing and retaining consumer loyalty.

In the present case, valuing the intangible assets of the firm from the higher margins earned on individual products would amount to merely arithmetic aggregating as all the products are retailed under the same brand name with sub-brand variants. However, it provides the company with options to launch newer products or variants directly at the higher price points. A step further, if the company were to decide to make a public offering of equity shares, the superior valuation of the firm could help offer the shares to new shareholders at a premium.

Each of these scenarios is conditioned by the longevity of the differentiation. Assuming that if a rival player in the industry were to commence planning for 100 % energy self-sufficiency at the present, it could take fewer than 4 years to implement and commission all the components. That would provide Wyke with a 3- to 4-year window to recoup its green energy investments. The instructor could extend the discussion to cover various other scenarios involving time windows, risks involved in the progression of events, higher costs of project debt etc.

The instructor could guide the students to evaluate breakeven value in a number of small steps as laid out here-in-below:

Financial Model

1. Table 1—Returns on assets and returns on equity earned by the company.
2. Table 2—Projected net margins corresponding to a target return on equity of 16 %, holding revenues constant: First polar case of all gains accruing from efficiency enhancement.
3. Table 3—Projected returns on assets corresponding to a target return on equity of 16 %, holding revenues constant.
4. Table 4—Projected revenues from the sale of the company's products corresponding to a target return on equity of 16 % (ignoring improvements in operational efficiency): Second polar case with all gains accruing from premium pricing alone.
5. Table 5—Year-on-year price increase for 3 years in keeping with projected total revenues computed in Table 4.

The cash-flow chart for the project would need to be drawn-up one step at a time.

1. Returns on assets and returns on equity

	31-03-2013	31-03-2012	31-03-2011	31-03-2010	31-03-2009
Profit after tax	1,817,000	1,670,000	931,000	−573,000	1,017,000
Total assets	**43,868,000**	**38,017,000**	**33,722,000**	**38,167,000**	**42,427,000**
average assets	40,942,500	35,869,500	35,944,500	40,297,000	
Return on assets (%)	4.1420	4.3928	2.7608	−1.5013	
Shareholder funds	**13,950,000**	**12,442,000**	**11,056,000**	**10,281,000**	**11,387,000**
Average equity	13,196,000	11,749,000	10,668,500	10,834,000	
Return on equity (%)	13.769	14.214	8.727	−5.289	

Data source http://companycheck.co.uk/company/00751654/WYKE-FARMS-LTD/financial-accounts#company-accounts
All figures in GBP

2. Scenario: Assuming no change in pricing or quantity sold (constant revenues) and 12 % year-on-year increase in shareholders funds, return on equity targeted at 16 % entirely from improved operational efficiency: net margin of almost 5 % earned in y3

	Projection			Actual		
	31-03-2016	31-03-2015	31-03-2014	31-03-2013	31-03-2012	31-03-2011
Turnover	59,820,000	59,820,000	59,820,000	59,820,000	59,260,000	59,075,000
Profit after tax	2,967,810.048	2,649,830.4	2,365,920	1,817,000	1,670,000	931,000
Net margin (%)	4.96	4.43	3.96	3.04	2.82	1.58
Shareholder funds	**19,598,745.6**	**17,498,880**	**15,624,000**	**13,950,000**	**12,442,000**	**11,056,000**
Average equity	18,548,812.8	16,561,440	14,787,000	13,196,000	11,749,000	10,668,500
Return on equity (%)	16.00	16.00	16.00	13.769	14.214	8.727

All figures in GBP

3. Scenario: Assuming no change in pricing or quantity sold (constant revenues)
   and 12 % year-on-year increase in shareholders funds, return on equity targeted
   at 16 % entirely from improved operational efficiency: Total assets remain
   constant at March 2013 levels, then 6.76 % return on assets earned in y3

	Projection			Actual		
	31-03-2016	31-03-2015	31-03-2014	31-03-2013	31-03-2012	31-03-2011
Turnover	59,820,000	59,820,000	59,820,000	59,820,000	59,260,000	59,075,000
Profit after tax	2,967,810	2,649,830	2,365,920	1,817,000	1,670,000	931,000
Net margin (%)	4.96	4.43	3.96	3.04	2.82	1.58
Total assets	**43,868,000**	**43,868,000**	**43,868,000**	**43,868,000**	**38,017,000**	**33,722,000**
Average assets	43,868,000	43,868,000	43,868,000	40,942,500	35,869,500	35,944,500
Return on assets (%)	6.7653	6.0405	5.3933	4.1420	4.3928	2.7608

All figures in GBP

4. Scenario: Assuming return on equity targeted at 16 % from improved pricing:
   Net margins of 3.96, 4.43, and 4.96 %, respectively, earned from improved
   revenues: with 4.96 % return on assets and 16 % return on equity earned in y3

	Projection			Actual		
	31-03-2016	31-03-2015	31-03-2014	31-03-2013	31-03-2012	31-03-2011
Turnover	60968544	60,651,498	60,370,344	59,820,000	59,260,000	59,075,000
Profit after tax	1,817,000	1,817,000	1,817,000	1,817,000	1,670,000	931,000
Net margin (%)	4.96	4.43	3.96	3.04	2.82	1.58
Total assets	**43,868,000**	**43,868,000**	**43,868,000**	**43,868,000**	**38,017,000**	**33,722,000**
Average assets	43,868,000	43,868,000	43,868,000	40,942,500	35,869,500	
RoA (%)	4.1420	4.1420	4.1420	4.1420	4.3928	2.7608
Shareholder funds	**19,598,745.6**	**17,498,880**	**15,624,000**	**13,950,000**	**12,442,000**	**11,056,000**

(continued)

(continued)

	Projection			Actual		
	31-03-2016	31-03-2015	31-03-2014	31-03-2013	31-03-2012	31-03-2011
Average equity	18,548,812.8	16,561,440	14,787,000	13,196,000	11,749,000	0.02,760,809
RoE (%)	16.00	16.00	16.00	13.769	14.214	

All figures in GBP

5. Scenario: Assuming return on equity targeted at 16 % from improved pricing: Net margins of 3.96, 4.43, and 4.96 %, respectively, earned from improved revenues: with 4.96 % return on assets and 16 % return on equity earned in y3: product pricing at 14,000 t per annum of output

	Projection			Actual		
	31-03-2016	31-03-2015	31-03-2014	31-03-2013	31-03-2012	31-03-2011
Annual sales (tonne)	14,000	14,000	14,000	14,000	14,000	14,000
Revenue from sales	60,968,544	60,651,498	60,370,344	59,820,000	59,260,000	59,075,000
Unit price (tonne)	4,354.90	4,332.25	4,312.17	4,272.86	4,232.86	4,219.64
Unit price (kg)	4.355	4.332	4.312	4.273	4.233	4.220
Unit price (% annual increase)	0.523	0.466	0.920	0.945	0.313	

References

Appleyard D (2013) Lush grass becomes lush green for UK biogas project. Renew Energy World Mag 56–59

Armenian Development Agency (2012) Cheese markets: cheese market research for Russian, Georgian, European and Arab Countries, p 63 of 139

Gray J (2013) Making Cheddar: personality and profitability in the West Country. The Manufacturer, 29 July 2013. http://www.themanufacturer.com/articles/making-cheddar/. Accessed 24 Mar 2014

Stones M (2014) Wyke farms hits key green energy target, foodmanufacture.co.uk, William Reed Business Media Limited. http://www.foodmanufacture.co.uk/Manufacturing/Wyke-Farms-hits-key-green-energy-target. Accessed 25 Mar 2014

Chapter 13
Norfolk Council Energy from Waste Project: Unfavorable Currents

Walk Away Price

> *I believe that this proposal has been subjected to the most intensive scrutiny, by the Planning Process, the Public, the Environment Agency and DEFRA, which has approved the largest grant Norfolk has received for a single project.*
> —Bill Borrett, Cabinet Member for Environment and Waste, June 2012

Introduction

The UK Government's Department for Environment, Food and Rural Affairs (Defra) decided to withdraw its 25 year—GBP 169 million infrastructure credit (formerly private finance initiative credits) from the 4.8 ha/GBP 500 million/ 268,000 tonne-per-year Willows Power and Recycling Center, energy from waste (EFW) facility near Kings Lynn, Norfolk (Messenger 2014b). An artist's impression of the completed project is depicted in Fig. 13.1.

The promissory note providing an unconditional guarantee from the treasury to disburse the grant in installments, upon successful commissioning of the plant, was awarded in January 2012, despite local opposition, which at that time was perceived to be "not, in itself, unusual in major energy from waste projects" (Waste Management World 2014a).

Willows Power & Recycling Center (WPRC) in Saddlebow, near King's Lynn, had received planning consent in June 2012, and the Norfolk County Council had achieved project financial closure. The construction contract was awarded to Cory Environmental of the UK and US Waste-to-Energy specialist Wheelabrator Technologies, a subsidiary of Waste Management, who in turn appointed Hitachi Zosen Inova as the engineering, procurement, and construction (EPC) contractors. The project was slated to:

© Springer India 2015
S. Sunderasan, *Cleaner-Energy Investments*,
DOI 10.1007/978-81-322-2062-6_13

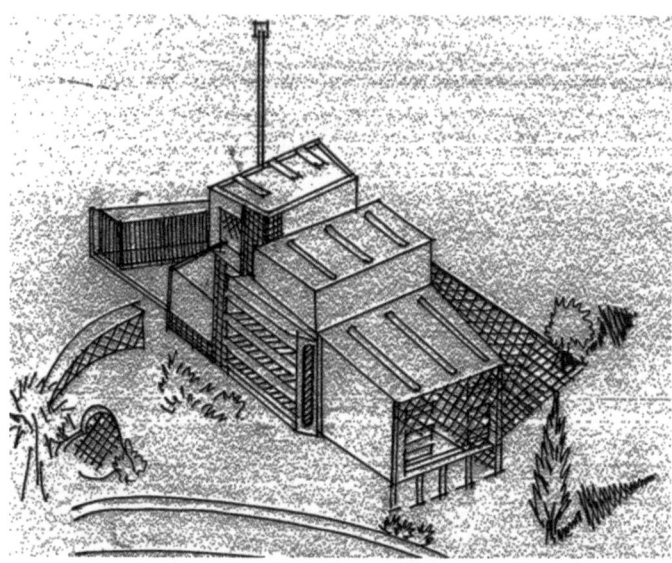

Fig. 13.1 Artist's impression of the Willows Power and Recycling Centre; Author's sketch

- Create over 1,000 jobs during construction
- Divert waste from landfills and achieve GBP 8 million savings per annum in waste disposal costs for the county
- Treat municipal (170,000 t) and some commercial and industrial wastes (98,000 t)
- Help reduce CO_2 emissions equivalent to taking 30,000 cars off the roads (assuming average car CO_2 emissions of 172.8 g/km and an average car running 8,340 miles a year), net of emissions caused during construction and of flue gases from waste incineration
- Recycle 55,000 t of metals and aggregates each year
- The 24-MW power plant was to generate 160.8 GWh (plant load factor of about 76.50 %) of electricity per year, equivalent to the consumption of 36,000 homes at an average of 4,478 kWh per annum per home (Messenger 2014a).

Schematic representation of the project is shown in Fig. 13.2 (*Figure credit* www.norfolk.gov.uk).

However, the Cory–Wheelabrator project received strong objections from local residents and politicians. The Secretary of State for Communities and Local Governments then "called in" the approval granted, citing "substantial regional and national controversy," this despite the stated policy of being "very selective" in calling planning applications in Waste Management World (2014b). In response, the consortium expressed disappointment over the withdrawal of consent for the Willows Power & Recycling Center and promised to "engage positively" to demonstrate the merits of their application.

Energy from waste and recycling plant

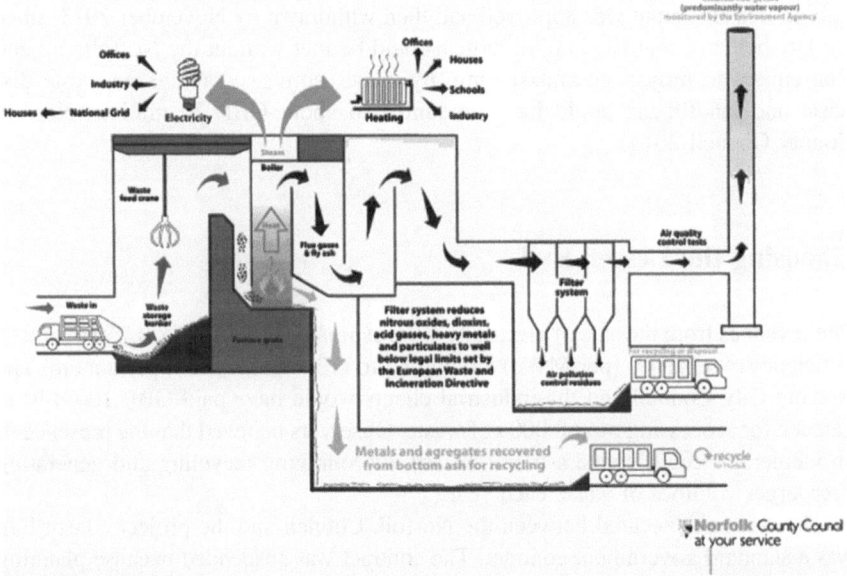

Fig. 13.2 Schematic representation of the Norfolk energy from waste project; *Figure courtesy* Norfolk County Council

Historical Background

Since 2006, the Norfolk councils had worked toward reducing, reusing, and recycling municipal and household waste with a view to minimizing material sent to landfills and more significantly the landfill tax expense. About half of all household waste was recycled by the year 2012–2013, and the residual waste was sent to the energy from waste facility in Kent.

The county found that (i) landfills were environmentally unsustainable, (ii) approved landfill sites were rapidly filling up, (iii) landfill operator gate fee and government taxes were rising rapidly (estimated to rise to GBP 80 per tonne aggregating to GBP 16.50 million by April 2014), (iv) missing EU-imposed targets for biodegradable municipal wastes reaching landfills in year 2020 would have meant additional fines of the order of GBP 150 per tonne, and (v) environmental and economic benefits could be derived from energy from waste plants and material recycling operations.

Consequently, as a part of a long-term solution to managing the residual waste, and after pursuing a long-drawn and rigorous procurement process, the Norfolk County Council's Cabinet had approved the Cory–Wheelabrator proposal to build and operate the Willows power and recycling center near King's Lynn. Environmental permit to operate the installation under the Environmental Permitting

(England and Wales) Regulations of 2010 was issued by the Environmental Agency in July 2012. The contractor was to invest about GBP 150 million. The GBP 169 million treasury grant was approved and then withdrawn by November 2013, since the UK believed that EU landfill targets could be met without the Norfolk project. Consequent to project commissioning, by some estimates, savings on waste disposal and landfill tax could have amounted to about GBP 20 million (Norfolk County Council 2014).

Choosing the Lesser Evil

The revenues from the sale of electricity were estimated at 24 MW × 8,760 h × 60 % (efficiency) × GBP 60 (per MWh) amounting to GBP 7.57 million per annum. The Norfolk City Council and the industrial clients would have paid GBP 100–140 in gate fee for processing of 268,000 t of waste. Observers believed that the presence of an incinerator could create a moral hazard, discouraging recycling and generating ever larger volumes of waste each year.

The contract executed between the Norfolk Council and the project consortium was a standard government contract. The contract was suspended because planning permission was first granted, and then withdrawn in the face of local opposition. Withdrawal from the contract would imply loss of the GBP 169 million grant approved by the treasury. Worse, the planning failure would also trigger a compensation payment to the contractor, due to be paid within 40 days of termination. The level of compensation was capped at GBP 20.30 million. Additionally, the council was to reimburse public inquiry costs of between GBP 1.50 and 2.00 million and losses arising due to drifts in exchange rates and interest rates of about GBP 11.00 million (Waste Management World 2014c).

In summary, the project was projected to lower costs and tax expenses associated with waste disposal, relative to landfills, and rejecting the project would have triggered an obligation for the county to recompense the contractor to the extent of GBP 30 million, which was likely to "effectively bankrupt the council" (Messenger 2014c).

By end October 2013, despite the withdrawal of the treasury grant, the Norfolk Council chose to go ahead with project implementation and to update contract dates for the project to address the delay in securing planning permission (Holder 2014).

In January 2014, two members of the British Parliament, Henry Bellingham and Elizabeth Truss, opined that at GBP 75–85 per tonne, shipping waste to Amsterdam for disposal would prove economically more attractive when compared to GBP 105 at the Norfolk, Willows Power Plant. It was also reported that the national average waste disposal cost was approximately GBP 78 per tonne (Grimmer 2014). Progressively, the debates involving the project were driven by political considerations rather than environmental and technical arguments.

Project Data

	Quantity/estimate	Unit charge (GBP) (Edwards 2014)
Revenues		
Norfolk City Council—waste disposal	168,000 t	140.00
Third-party waste	100,000 t	140.00
Sale of electricity	107,000 MWh	60.00
Material recovery	Estimate per annum	8.0 million
Costs		
Operating costs including landfill charge	Estimate per annum	15.0–18.0 million
Project development cost	Estimate	5.0 million
Cost of capital	Assumed	13.5 %

Sequence of Project-Related Events

http://www.willowsprc.co.uk/latestnews.php, last accessed February 14, 2014.

Date	Event
October 29, 2013	NCC's Cabinet accepts revised project plan
October 18, 2013	DEFRA withdraws waste infrastructure credits
October 16, 2013	ASA dismisses latest complaints about Willows advertising
May 23, 2013	Cory Wheelabrator statement—Willows proposal achieves energy recovery status
May 17, 2013	Cory Wheelabrator—public inquiry closing statement
	Cory Wheelabrator—public inquiry opening statement
February 26, 2013	Cory Wheelabrator—Public inquiry closing submissions
August 30, 2012	DCLG to call in planning application for the Willows Power and Recycling Centre
July 31, 2012	Environmental permit is granted by the Environment Agency for the Willows Power
June 29, 2012	Norfolk County Council's Planning (Regulatory) Committee approves the planning application for the Willows Power and Recycling Centre

(continued)

(continued)

Date	Event
June 19, 2012	Norfolk County Council publish officer's report into proposed Willows Power and Recycling Centre
May 14, 2012	Clarification on connections to the national grid electricity network
March 12, 2012	Cory Wheelabrator Statement to King's Lynn and West Norfolk Borough Council's Planning Committee
March 9, 2012	NHS Norfolk and Waveney maintains its broad support for the Willows Power and Recycling Centre
March 1, 2012	Natural England withdraws its objection
February 27, 2012	Environment Agency withdraws its objection
February 8, 2012	Cory Wheelabrator Consortium signs contract with Norfolk County Council
February 8, 2012	ASA dismisses complaints about Willows advertising
January 30, 2012	Norfolk's business leaders turn out in force to find out about jobs and supplier opportunities
January 24, 2012	Major contract awarded to Norfolk firm
January 18, 2012	Norfolk County Council waste proposal secures Government support worth £169 million
November 17, 2011	City of London achieves zero to landfill status
November 11, 2011	Norfolk County Council responds to Caroline Spelman's possible withdrawal of PFI credits
July 25, 2011	Cory Wheelabrator Statement to King's Lynn and West Norfolk Borough Council's Development Board
June 27, 2011	Second Willows Power and Recycling Centre Newsletter
June 15, 2011	Planning application submitted
March 21, 2011	Factsheet on energy from waste published
January 19, 2011	Willows Power and Recycling Centre business open day press release
December 1, 2010	Willows Power and Recycling Centre Newsletter
October 27, 2010	Cory Environmental and Wheelabrator Technologies recommended as preferred bidder for Norfolk PFI
October 27, 2010	Preferred bidder recommended for Norfolk's residual waste disposal PFI project Norfolk County Council press release (http://www.norfolk.gov.uk/news/NCC085349)

Teaching Note

Case Synopsis

The EfW project had received planning permission and treasury grant support after a long-drawn procurement process and after a thorough evaluation of all applications. The consortium of Cory Environmental of the UK and Wheelabrator of the USA was selected to construct and operate the Willows Power and Recycling unit outside of King's Lynn in Norfolk County. The project had faced political and social challenges all through. Initially, the Council members considered the protests (by as many as 65,000 people forming 94 % of the population) as being routine, but eventually, the planning permission was called in. By October 2013, the treasury decided to cancel the grant citing the country's ability to meet EU recycling targets without having to build the Norfolk plant.

The council's contract with the consortium, however, provided for a compensation to make good losses from project preparation, public hearing, financial closure, and other upfront investments. It was estimated that if the council were to walk away from the project, the compensation would amount to about GBP 30 million, effectively driving the county council into bankruptcy. This left the council with few options. It was therefore decided that the project would be implemented even without the treasury grant. However, in the revised situation, margins were relatively slim and an eventual default by the council after plant construction would lead to greater losses for the investor consortium. Middle ground had to be explored to find an amicable settlement.

Case Question

The case is intended to highlight the political and social currents at play when public–private partnership contracts and federal grants are involved. The options before the consortium would be to walk away with the compensation amounts or to implement the project and maximize shareholder returns. The authorities would need to plan a financing package that would keep the investors interested and yet not drive the Council itself into financial difficulty.

Teaching Objectives

- Enabling students appreciate the political and social challenges embedded when public projects are implemented and operated by private sector entities.
- Benchmark alternative scenarios against the business as usual (BAU) scenario.
- Track the sequence of activities and map the gradual degeneration of the debate from environmental issues to political rhetoric.

- To help restructure the investment to balance investor returns with client solvency.
- Make suitable recommendations for this and future projects.

Case Objectives and Use

The crux of the case is evaluating the sources of risk and appropriately managing them proactively. The project is not projected to absorb technology risk as waste incineration has been employed elsewhere. The retrieval of metals and other residues of commercial value offset the risk from the possible lowering of tipping (gate) fee by the council or third-party aggregators. The case seeks inputs aimed at balancing sources of revenues with public objections: to try and reduce the gate fee for waste while earning returns from the sale of electricity, which is a by-product from the process. However, this would have to be balanced against the potential moral hazard relating to higher quantities of waste generated from lower recycling rates.

The project would need to balance between the gate fee and the quantity of waste received at the plant gate while ensuring solvency of the council itself. The electricity could be exported to other pockets within the county or outside where it could be in demand.

Teaching Plan

A technically qualified consortium develops a project proposal and receives all approvals, which are eventually revoked. The project's environmental benefits are apparently no longer required, as the UK expected to meet the EU's 2020 targets without having to implement the project.

In other words, the project would have to be viewed independent of its environmental characteristics: as a process that took municipal waste in for a fee, and generated electricity and residual metals and materials. The process had to prove its viability and sustain commercial operations over the period of the concession.

The instructor could help evolve scenarios for different levels of gate fee (tipping fee), landfill taxes, electricity tariffs, etc., to analyze the project's financial formulation. The analyses should take the following into account:

1. Agencies that would constitute the project's competitors: power plants, anaerobic digestors/compost plants, etc., and other cleaner energy technologies that would help the UK meet its environmental commitments.
2. Agencies that would constitute the project's complementors: waste aggregation, sorting and transportation operators, utility network operators, waste metal recyclers, and brick and construction material distributors.
3. The financial model reworked to factor in a delay in project implementation.

4. The impact of a delay in implementation on the attractiveness of the investment: higher interest costs, higher start-up and preoperative costs, higher landfill costs, possibly higher revenues from the sale of electricity, etc.
5. Strategies recommended to make the project more viable.

What Happened Next

As of mid-January 2014, the Borough Council of King's Lynn & West Norfolk announced that a residual waste project proposed by Chester-based *Material Works Limited* had received GBP 100 million in funding from a consortium of "Asian business people." A planning application was to be developed to build the facility in west Norfolk. The plant was to receive and process 35,000 t of "black bag waste" collected by the Borough Council, 5,000 t of food waste, and an additional 35,000 t of industrial and commercial waste (Reece 2014). The plant would remove recyclable material, extract methane for electricity generation, and generate a sterilized and homogenized mixture *Omnicite* and then a polymer alloy *Rexylon*. Gaining planning permission for the proposed project was viewed as a major hurdle, though. Dr. Colin Church, Departmental Director at the UK's Department of Environment, Food and Rural Affairs, in early March 2014, said that greater importance would be placed on producing heat from waste as opposed to generating electricity and that the "E" in EfW did not stand for electricity (Bayar 2014). By mid-March 2014, UK's Chancellor of the Exchequer presented the annual budget and announced that with effect from April 2015, the standard and lower rates of landfill tax would increase in line with inflation (Messenger 2014d).

References

Bayar T (2014) UK to focus on CHP in waste-to-energy schemes. Cogeneration and Onsite Power, 3 Mar 2014. http://www.cospp.com/articles/2014/03/uk-to-focus-on-chp-in-waste-to-energy-schemes.html. Accessed 20 Mar 2014

Edwards C (2014) Broadly overlapping with. The Proposed Incinerator at King's Lynn. Report of February 2011. http://www.farmerscampaign.org/pages/pfi.html. Accessed 14 Feb 2014

Grimmer D (2014) MPs call for Norfolk Country Council to scrap incinerator. http://www.edp24.co.uk/news/politics/mps_call_for_norfolk_county_council_to_scrap_incinerator_1_3245125?ot=archant.PrintFriendlyPageLayout.ot. Accessed 12 Feb 2014

Holder M (2014) Norfolk considering JR over defra funding loss. 29 Oct 2013. http://www.letsrecycle.com/news/latest-news/councils/norfolk-considering-jr-over-defra-funding-loss. Accessed 1 Feb 2014

Messenger B (2014a) Waste to energy advert not misleading say UK's advertising standards authority. Waste Management World, 18 Oct 2013. http://www.waste-management-world.com/articles/2013/10/waste-to-energy-advert-not-misleading-say-uk-s-advertising-standards-authority.html. Accessed 1 Feb 2014

Messenger B (2014b) Norfolk waste to energy project loses £169 M government funding. Waste Management World, 21 Oct 2013. http://www.waste-management-world.com/articles/2013/10/norfolk-waste-to-energy-project-loses-169m-government-funding.html. Accessed 1 Feb 2014

Messenger B (2014c) Revisions to Norfolk's Cory Wheelabrator waste to energy contract approved. Waste Management World, 29 Oct 2013. http://www.waste-management-world.com/articles/2013/10/revisions-to-norfolk-s-cory-wheelabrator-waste-to-energy-contract-approved.html. Accessed 4 Feb 2014

Messenger B (2014d) Reaction as UK budget sees little change to landfill tax but funding to tackle waste crime. Waste Management World, 19 Mar 2014. http://www.waste-management-world.com/articles/2014/03/reaction-as-uk-budget-sees-little-change-to-landfill-tax-but-funding-to-tackle-waste-crime.html. Accessed 25 Mar 2014

Norfolk County Council (2014) Future of household waste in Norfolk. http://www.norfolk.gov.uk/Environment/Waste_and_recycling/Waste_collection_and_disposal/Future_of_household_waste_in_Norfolk/index.htm. Accessed 4 Feb 2014

Reece A (2014) Material works secures £100 million funding. Resource Magazine, 16 Jan 2014. http://www.resource.co/government/article/material-works-secures-%C2%A3100-million-funding. Accessed 14 Feb 2014

Waste Management World (2014a) Norfolk waste to energy facility awarded £91 Million WI Grant. 19 Jan 2012. http://www.waste-management-world.com/articles/2012/01/norfolk-waste-to-energy-facility-awarded-p91-million-wi-grant.html. Accessed 3 Feb 2014

Waste Management World (2014b) Norfolk waste to energy planning application in a pickle. 31 Aug 2012. http://www.waste-management-world.com/articles/2012/08/norfolk-waste-to-energy-planning-application-in-a-pickle.html. Accessed 4 Feb 2014

Waste Management World (2014c) Cancelling 24 MW waste-to-energy plant could cost Norfolk £90 M. 29 May 2013. http://www.waste-management-world.com/articles/2013/05/cancelling-24-mw-waste-to-energy-plant-could-cost-norfolk-p90m.html. Accessed 4 Feb 2014

Chapter 14
Offshore Wind Turbine Foundations: Staying Afloat in Deepwaters

Risk–Reward Relationships in Technology Development

Japan is one of the most exciting energy markets in the world today. The closure of all its nuclear reactors is a challenge that requires an effort out of the ordinary, and new renewables technology needs to play a large part in the solution.
—Johan Sandberg, RECharge News, 3 February 2014

Floating offshore wind is a symbol of a [green] future.
—Yuhei Sato, Honourable Governor of Fukushima, Japan

Introduction

Longer working hours, higher plant efficiency factors, and superior reliability had brought offshore wind energy systems to the threshold of qualifying as base-load generation, alongside centralized conventional power plants. Sustained gains in deployment and maintenance efficiency promised to guarantee security of supply, grid stabilization, and lower total costs within future energy system combinations (Fig. 14.1).

Given favorable seabed conditions and the most reliable wind resources in the world, the North Sea alone was estimated to produce energy amounting to 400 % of European demand (Northland Power Inc 2014). Additionally, if offshore wind energy capacity were to constitute a significant proportion of the total generation mix, it could reduce the need for backup storage capacity as well (Garus 2014). In contemporary developments, Spanish wind turbine major Gamesa and French conglomerate Areva had floated an equally owned joint venture to pool resources in offshore wind energy development and to exploit the potentially large European market. Among other things, the venture was to accelerate the development of the next generation of turbines, commencing with an 8-MW model (Williams 2014).

© Springer India 2015
S. Sunderasan, *Cleaner-Energy Investments*,
DOI 10.1007/978-81-322-2062-6_14

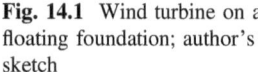

Fig. 14.1 Wind turbine on a floating foundation; author's sketch

Floating Offshore Wind Farms

In a symbolic evolution, in the month of November 2013, a floating wind turbine had commenced operations 20 km from the stricken nuclear power station in Japan's Fukushima prefecture. The pilot project was funded by the Japanese government, and the development was led by Marubeni Corporation. The 2-MW turbine was supplied by Hitachi Limited and was supported by a 66-kV floating substation and extra-high-voltage undersea cables. The Japanese Ministry of Economy, Trade and Industry (METI) had proposed expanding floating offshore wind energy generation capacity to 1,000 MW, and toward this end, Mitsubishi Heavy Industries Limited was to install two 7-MW pilot units. At 16 MW of installed capacity, by 2014, the experimental setup was slated to be the world's largest floating offshore wind farm. In areas that were too deep for traditional towers that would be anchored to the seabed, the technology involved attaching turbines to floating structures, thus opening up large deep-ocean areas for clean energy production (Rose 2014).

Marubeni was entrusted with the project management responsibility for the USD 189 million (JPY 18.80 billion) project. Ship building major, Japan Marine United, and the University of Tokyo had built a floating platform for the floating substation. Mitsui Corporation had built the semi-submersible foundation for the turbine. Concurrently, METI also set up a panel of experts to assess the merits of a premium feed-in tariff that would take the considerably higher costs of offshore wind into account. As of end-2013, both onshore and offshore wind energy installations received identical tariffs of 23.10 Japanese Yen/kWh ($\sim$ USD 0.232/kWh), for a 20-year period. The preferential tariff for offshore wind energy generation was scheduled to come into effect by April 1, 2014 (Publicover 2014).

In depths greater than 50 m, foundation costs were found to render offshore wind generation projects prohibitively expensive. The technical solution involved lowering the center of gravity below the center of buoyancy to stabilize the structure and then mooring it by catenary lines. Norwegian energy company Statoil Hydro had installed the world's first deepwater, floating (2.3 MW) turbine in June 2009. The 93.0-m-diameter rotor turbine sourced from Siemens of Germany and the 120-m-high tower were mounted on a 117-m-long, 3,000-t floating cylinder, inspired by similar applications in the oil and gas industry. The assembly was towed offshore into 220 m deepwater, 10 km from the shore. The electricity generated was fed to the Norwegian grid through submarine power transmission cables (Stiesdal 2009). The project costs about USD 62 million to build and install and was estimated to generate 9,000 MWh of electricity each year. The turbine generated 7,300 MWh of electricity in the year 2010, representing a plant efficiency factor of about 36 %. The schematic and an artist's impression of the design are presented in Fig. 14.2.

Seattle-based *Principle Power* installed a 2-MW *Vestas* make turbine on a floating foundation 3 miles off the Portugal coast: The wind farm was expected to grow to 150 MW in installed capacity. Principle had also received approval from the US Department of the Interior to build a 30-MW floating wind farm off the Oregon coast, the country's first in the Pacific Ocean. The USD 200 million farm would comprise 5 numbers of 6-MW Siemens turbines, 15 miles off the coast, and was expected to commence operations by the year 2017.

In late 2013, French wind farm developer *Quadran* partnered with foundation specialist *Ideol* to develop 500 MW of offshore wind power capacity by the year 2020, among others, using a custom-designed floating square concrete platform along the lines of the design, which was also to provide adequate surface area for O&M at the base of the turbines, without having to bring a vessel alongside the turbine. The concept included a damping pool to minimize dynamic stresses that could be induced and a base that could function as a helicopter landing pad.

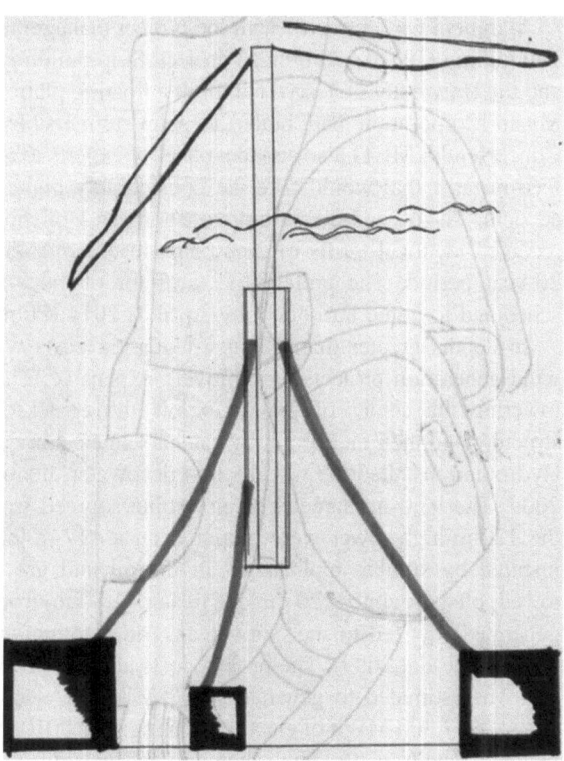

Fig. 14.2 Hywind floating turbine (Norway): schematic image author's sketch

The Offshore Opportunity

Deeper oceans and locations further away from the shore offered flexibility for installing turbines bypassing shipping lanes, fishing banks, bird migration paths, defense testing sites, and areas used for recreational pursuits. Offshore wind resource potential in the USA, China and Japan and other potential markets was recorded in waters deeper than 30 m. The vast offshore spaces off the Baltic states of Estonia, Latvia, and Lithuania were said to enjoy robust wind resources and favorable conditions allowing for lower cost (by about USD 1.40 m per MW). While project development in Lithuania had been slower than projected and despite the Estonian feed-in tariff (FiT) per kWh being progressively lowered from USD 1.31 (2012) and USD 1.2 (2013) to USD 1.10 (2014) (Jegelevicius 2014), developers had remained optimistic about the region's potential.

Most commercial offshore wind farm installation activity was witnessed off the European shores, especially off the UK. In calendar year 2013, a record-setting 1,567 MW of new offshore wind energy generation capacity was added in Europe.

With 418 new machines hoisted into place, this capacity represented a 33 % increase over the year 2012. 72 % of the new installations were located in the North Sea, followed by 22 % in the Baltic Sea and 6 % in the Atlantic Ocean. The cumulative installed offshore wind power, consequently, rose to 6,562 MW, meeting a shade under 1 % of the European Union's electric power requirements. It was estimated that some 22,000 MW of projects had received planning consent for implementation. Developing the massive wind energy potential in Europe was seen as a means to achieving green growth, generating employment, demonstrating technology leadership, and achieving CO_2 reductions. In year 2013, as with year 2012, Siemens, Dong Energy, and Bladt were market leaders in turbines, project development, and substructure, respectively (Azau 2014).

Figure 14.3 alongside shows the world's largest offshore wind farm, the 175 turbine—630-MW London Array as shot from *Landsat 8* satellite—the white dots being the wind turbines (figure by, NASA earth observatory). The wind farm is located 20 km from the Kent and Essex coasts and was spread over an area of 104 km^2.

The individual turbines were separated by distances ranging from 640 to 1,200 m and stood 147 m tall. Cables beneath the seafloor connected the turbines and transmitted power to two offshore substations and one onshore station. The farm had commenced operations in April 2013. The wind turbines were hosted on natural sandbanks, with maximum water depths of about 25 m. The location of the farm was optimized to be proximate to onshore power grids and simultaneously to minimize disruptions to shipping lanes. The wind farm was expected to grow to a spread of 246 km^2 and to reduce 900,000 t in CO_2 emissions (Chow 2014).

Fig. 14.3 *Landsat 8* satellite image of the 630-MW London Array Offshore Wind Farm; figure credit: London Array Limited, reproduced with permission

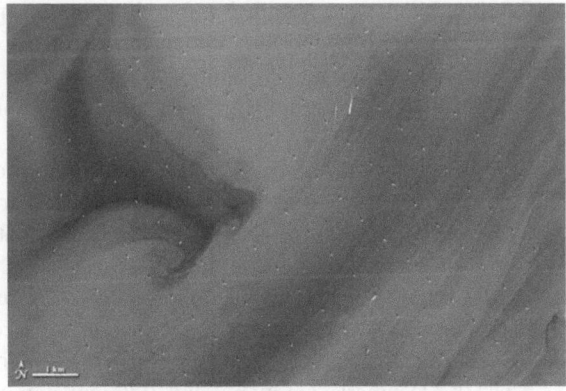

According to a report by the European Wind Energy Association (EWEA), as of end of December 2013,

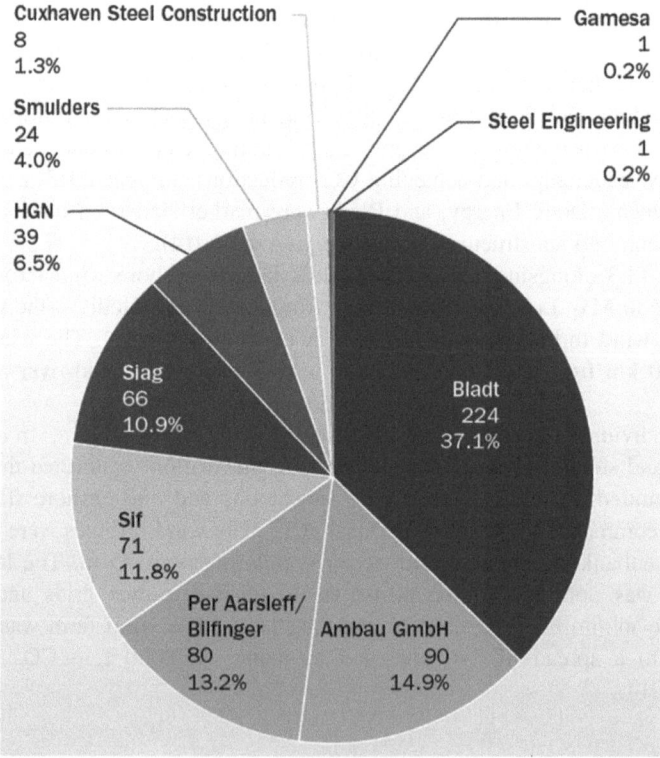

- 2,080 turbines were installed and grid-connected, making a cumulative total of 6,562 MW, in 69 wind farms in eleven European countries.
- In 2013 alone, 79 % (490 units) of substructures were monopiles, 14 % (87 units) tripods, 6 % (39 units) jackets, and 1 % (8 units) tripiles and 1 unit gravity foundation.
- Overall, of the 2,474 substructures, 76 % were monopiles, 12 % were gravity type, and 5 % each were jacket foundations and tripods.
- There were also two full-scale grid-connected floating turbines and two down-scaled prototypes.
- Bladt, with 224 foundations, followed by Ambau with 90 and Per Aarsleff/ Bilfinger with 80 were among the leaders in a market dominated by a total of 11 players (pie chart/graph, Fig. 14.4 alongside: credit EWEA).
- The UK, Denmark, Germany, and Belgium dominated the European offshore surge with 733, 350, 240, and 192 MW, respectively.
- Interestingly, however, the weighted average depth of wind farm installations in 2013 was 20 m, shallower than the 22 m depths of year 2012, while the average distance to the shore was of a similar range (30 km vs. 29 in 2012) (EWEA 2014).

Fig. 14.4 Progressively, wind farm sizes, distance to shore, and water depths are likely to increase: farms under construction in year 2013

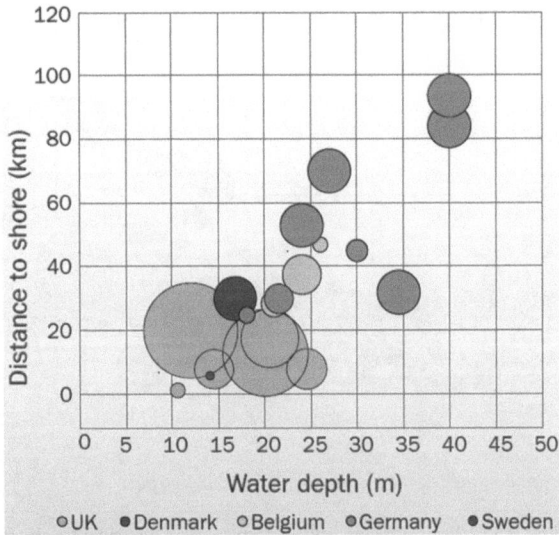

Tension-leg platforms, spar buoys, and semi-submersible floating platforms were all fastened to the seabed by cables. Floating offshore wind turbines were slated to be critical in exploiting the high-wind sites identified further away from the shore and in deepwaters and hostile marine environments. Fixed foundations as with nearshore installations would be unfeasible at such depths. Also, floating installations could be assembled onshore and towed out to deepwaters. Reliability of generation and supply would crucially hinge on balancing among turbine sizes, substructure design, grid connections, and control systems. Japan, Norway, and the USA represented emerging and sizable future markets (O' Donovan 2014).

Commercial floating wind farms were expected to benefit from the long years of experience gained from fixed-foundation wind power as well as from the progressive evolution in turbine and blade sizes. It was expected that infrastructure costs for deepwater turbines would not necessarily increase in proportion to energy generation from ever larger turbines. Consequently, as shown in Fig. 14.5, over the next few years, offshore wind farms were projected to be located further from the shore and at greater depths.

Floating foundations were known to impose additional and dynamic loads on the structure, and the impacts of such incremental loading on O&M requirements were yet unknown. In principle, however, floating turbine foundations were expected to reduce offshore O&M costs as grouting was eliminated and connections were simpler. Further, the floating assemblies could be towed back to shore in case a major overhaul was required (Deign 2014). The lower O&M costs combined with higher efficiency factors (of about 40 %) were slated to more than justify the higher investment costs.

The capital costs of the wind turbines themselves were not expected to be much higher, relative to shallow water installations. The additional costs of the floating

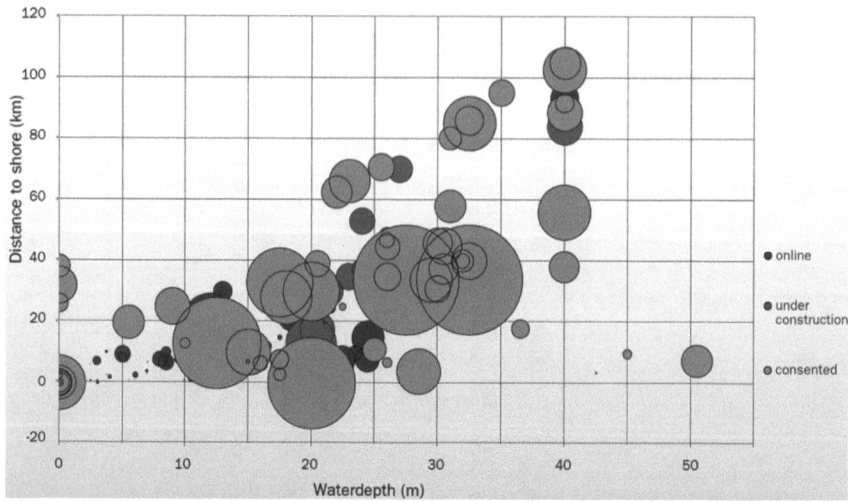

Fig. 14.5 Progressively, wind farm sizes, distance to shore, and water depths are likely to increase: farms under construction and consented; figure credit EWEA

structure and power distribution system were slated to be offset by higher productivity from the reliable wind resource further off. While onshore wind turbines were estimated to cost between USD 1.0 and 1.5 million per installed MW and shallow water turbines were estimated to cost between 2.4 and 3.0 USD million, the practical feasibility and per-unit economics of deepwater floating offshore turbines were yet to be clearly established. Researchers observed that prototype floating deep-offshore turbines were being deployed at about USD 6.0 per MWh and that the price would be lowered to about USD 4.0–5.0 with experience and learning (Doom 2014).

Teaching Note

Case Synopsis

Deepwater and further offshore wind installations were seen as the next major market opportunity after nearshore and shallow sites were almost saturated. With objections to nearshore wind farms from recreational users of the shorefronts, and given the risks posed to migrating birds and possible obstructions to shipping lanes, deep offshore wind was seen as a safe bet. Anchoring turbines directly to the seabed at water depths greater than 50 m was found to be prohibitively expensive.

Numerous efforts had been made to adapt technologies from the offshore rig structures employed by the oil and gas industry. A few customized designs to suit the specifics of wind turbines were also being tested both at laboratory scale and on the field. No single design had emerged a winner as yet. The additional costs of moving further offshore and into greater depths were to be offset by the lower operations and maintenance (O&M) costs and by the higher efficiency factors (projected at 40 % or more). Further, the power output from deep-offshore wind farms was slated to be more predictable and stable, even comparable to base-load generators and hence would have avoided the costs of storage technologies.

The case provides for indications of the costs of alternative test facilities. The projects had been funded by the governments of Norway, of Japan, etc., and were yet to confirm commercial viability.

Case Focus

The case is intended to highlight the risk–reward relationship in new technology development. Additionally, it focuses on the involvement of the private sector in implementation of new technology demonstration projects. One or more of the candidate technologies could make it to commercial viability, but picking the winner at the design stage could be challenging.

Teaching Objectives

- Enabling students to keep the ultimate targets of commercial viability in mind while engaging in technology development projects.
- Drill down the impact of design development costs on the developer's financial standing as well as on the ultimate benefits derived by each turbine employing the technology.
- Balancing design development costs with the viability of deep-offshore instal-lation through the determination of appropriate royalty (or other) payment mechanisms.

- Develop scenarios for threshold costs that would ensure viability of the new technology relative to most plausible alternatives including onshore and near-shore wind energy.

Case Objectives and Use

The case highlights the uncertainties associated with technology development projects. It encourages course participants to keep ultimate commercial application of the technology in view at each stage of the development process. While it would be easy to get carried away by technical considerations, the pragmatism of the solution should hold center stage at all times in coming up with alternative means to achieving these effects.

The candidate designs are funded by governments as a means of encouraging risk taking and eventually with a view to making the solutions available to all market participants. When privately funded, the returns on the RD&D investments are earned by royalty or technology transfer payments. These payments hold the key to balancing between funding for technology development and ensuring viability of individual installations. Further, the offshore feed-in tariffs that were being designed by authorities in Japan, for instance, need to balance between higher first costs and lower O&M costs and higher productivity from deep-offshore installations. The case in effect presents an optimization problem for analysis and resolution.

Teaching Plan

Very frequently, environmentally sensitive technology, such as wind energy generation, has been at the receiving end of environmental activism. Wind farms have been charged with disrupting flight paths of birds, disturbing radar signals, or merely being unsightly hardware-obstructing views of the horizon. Over the years, onshore sites were first saturated, and progressively, nearshore sites were being occupied and exploited. The next frontier for expansion was deep-offshore and deepwater sites, which came with their own set of challenges.

The instructor could guide course participants to analyze the benefit–cost perspective of a design development project. This could then be coupled with an economic analysis of individual wind turbines—additional upfront costs, lower O&M charge, and higher output, with possibly higher revenues (or lower taxes) from carbon emission mitigation.

Course participants could discuss the merits of public investments into research and development with its allied approval processes, delays in procurement, etc., but with the fruits being made available to society as a whole (public goods), versus private investments that are more directed at a specific market opportunity, efficiently executed and monitored and then provided to industry for a fee. The externality benefits from research beyond immediate project boundaries and the limitations to patent protection and enforcement could be discussed.

Future Prospects

According to the Global Wind Energy Council (www.gwec.net), as of mid-2012, installed offshore wind farm capacity amounted to about 2 % of global installed electricity generation capacity, with Europe leading the way. Overall capacity growth of year 2013 installations relative to 2012 was at 12.5 %, with growth performance in 2014 slated to be better: 45 GW were added in 2012 and 35 GW in 2013, and projections indicated 45 GW additions for 2014 (Pischel 2014). Japan's very capable maritime industry was slated to help grow the country's deepwater installation base quite rapidly and to contribute significantly to the domestic energy supply, displacing the country's reliance on nuclear power.[1]

The offshore wind industry continued to be under pressure to bring costs down and significantly so. Deeper waters, difficult bottom conditions, and higher waves had contributed to driving costs higher. Higher construction costs, protecting equipment from salt spray, and managing grid connections to installations further offshore substations and turbines were all considered too expensive for commercial sustainability. The mass rollout of larger wind turbines further offshore was projected to lower unit costs and to attract non-recourse financing from mainstream investors.

What Followed

Dong Energy decided to place a bulk order for Vestas Wind Systems' 8 MW/224-m-high machine, for the UK Burbo Bank Extension Offshore Wind Farm project. This was in furtherance of its efforts to float fewer and larger turbines and hence to build fewer foundations and reduce overall costs by 30–40 %, thereby delivering energy at prices comparable with nearshore and onshore installations (Morales 2014). At about the same time, the proposed 240 MW expansion of the London Array Offshore Wind Farm was canceled owing to its potential impact on the red-throated diver, a bird species that spent winters in the area earmarked for the proposed farm (Bayar 2014). In March 2014, Japan announced feed-in tariffs of 36 Japanese yen a kilowatt-hour for power from offshore wind generators (Watanabe 2014).

---

[1] http://www.gwec.net/global-offshore-current-status-future-prospects/, last accessed 13 February 2014.

In May 2014, the US Department of Energy (DoE) picked three pilot offshore wind energy projects, each of whom would receive about USD 47 million in funding over a 4-year period through 2017 to support design, development, and deployment efforts. Of the three projects, the "Wind Float Pacific" project by *Principle Power*, 18 miles off the coast and in 1,000 foot deepwater incorporated a semi-submersible floating foundation, similar to the pilot deployed off the coast of Portugal. The Virginia Offshore Wind Technology Advanced Project entailed installing two units of 6-MW direct-drive turbines at a distance of 20 nautical miles offshore in 50 foot deepwaters, employing a hurricane-resilient, twisted-jacket foundation (Montgomery 2014).

References

Morales A (2014) Vestas's biggest turbine picked by Dong for UK offshore farm. Bloomberg 18 Feb 2014. http://www.renewableenergyworld.com/rea/news/article/2014/02/vestass-biggest-turbine-picked-by-dong-for-u-k-offshore-farm. Accessed 28 Mar 2014

Publicover B (2014) Japan set for Offshore FIT—Official. Rechargenews.com, 8 Jan 2014. http://www.rechargenews.com/wind/offshore/article1348341.ece. Accessed 18 Jan 2014

O' Donovan C (2014) Bright lights for tomorrow. Power Engineering International, Jan 2014, pp 20–23

Watanabe C (2014) Japan approves solar power tariff cut, sets offshore wind. Bloomberg, 25 Mar 2014

Rose C (2014) Floating offshore wind turbines could drive Japan's renewable energy future. RenewableEnergyWorld.com, 18 Nov 2013. http://www.renewableenergyworld.com/rea/blog/post/print/2013/11/floating-offshore-wind-turbines-could-drive-japans-renewable-energy-future. Accessed 18 Jan 2014

Chow D (2014) World's largest offshore wind-farm seen from space, 21 Jan 2014. http://www.space.com/24358-london-wind-farm-photo.html. Accessed 1 Feb 2014

Williams D (2014) Areva and Gamesa join forces to target offshore wind. Power Engineering International, 21 Jan 2014. http://www.powerengineeringint.com/articles/2014/01/areva-and-gamesa-join-forces-to-target-offshore-wind.html. Accessed 1 Feb 2013

EWEA (2014) The European offshore wind industry—key trends and statistics 2013, Report by the European Wind Energy Association, Jan 2014, pp 1–22

Stiesdal H (2009) Hywind: the World's first floating MW-scale wind turbine. Wind Directions, Dec 2009, p 52–53

Montgomery J (2014) DEE picks winners for US offshore wind development. Renewable Energy World, 7 May 2014, http://www.renewableenergyworld.com/rea/news/article/2014/05/doe-picks-winners-for-u-s-offshore-wind-development?cmpid=WNL-Friday-May9-2014. Accessed 9 May 2014

Deign J (2014) O&M considerations for floating turbines, 3 Feb 2014. http://social.windenergyupdate.com/operations-maintenance/om-considerations-floating-turbines?utm_source=e-brief0402&utm_medium=email&utm_content=0402&utm_campaign=WEU. Accessed 5 Feb 2014

Doom J (2014) Walt musial of the NREL quoted by Justin Doom. Floating wind farms venture further out to sea. Renewable Energy World, 4 Mar 2014. http://www.renewableenergyworld.com/rea/news/print/article/2014/03/floating-wind-farms-venture-farther-out-to-sea. Accessed 15 Mar 2014

Garus K (2014) Offshore wind energy provides economic benefits. http://www.sunwindenergy.com/node/72627, 26 Nov 2013, Picture courtesy www.fukushima-forward.jp. Accessed 18 Jan 2014

Jegelevicius L (2014) Baltics' Estonia ramps up wind power generation. Renewable Energy World, 10 Feb 2014. http://www.renewableenergyworld.com/rea/news/article/2014/02/baltics-estonia-ramps-up-wind-power-generation?cmpid=WNL-Wednesday-February12-2014. 12 Feb 2014

Northland Power Inc (2014) Northland power to up majority stake in 600 MW Gemini offshore wind project. PennEnergy, 30 Jan 2014. http://www.pennenergy.com/articles/pennenergy/2014/01/northland-power-to-up-stake-in-600-mw-gamini-offshore-wind-power-project.html. Accessed 3 Feb 2014

Azau S (2014) Record offshore wind figures conceal slow-down in new projects. Renewable Energy World, 29 Jan 2014. http://www.renewableenergyworld.com/rea/blog/post/2014/01/record-offshore-figures-conceal-slow-down-in-new-projects. Accessed 1 Feb 2014

Pischel T (2014) Global wind market: low growth in 2013. 13 Feb 2014. http://www.sunwindenergy.com/node/72972. Accessed 14 Feb 2014

Bayar T (2014) Expansion plans shelved for world's largest wind farm. http://www.powerengineeringint.com/articles/2014/02/expansion-plans-shelved-for-world-s-largest-offshore-wind-farm.html. Accessed 25 Mar 2014

Chapter 15
GoBiGas: Fueling the Biogas Movement

Keeping Greener Company

> *Gasification is not a very complicated technology, as long as you can get the fuel in and the ash out!*
> —Lars Waldheim, Waldheim Consulting, Sweden

> *... a particularly innovative project and it does not create a significant distortion of competition and trade between member states.*
> —Lars Waldheim, Waldheim Consulting, Sweden

Background

The 13-member municipalities of Sweden's Gothenburg (Swedish: Göteborg) region placed great emphasis on sustainable economic growth and high employment levels. Under 1 % of the energy required for district heating and cooling systems in the region was derived from oil. Over 80 % of the indoor ambient heat and hot water in the region was derived from refuse incineration, refinery and industry waste heat, electricity generation, and waste water treatment. In addition to being home to some of the most energy-efficient petroleum refineries in the world, the region was one of the world's largest producers of biogas, distributed through the natural gas grid. In 2008, the Business Region Goteborg (www.businessregiongoteborg.com) was awarded the Charles R. Imbrecht "Blue Sky Innovation Award" for encouraging the use of biomethane as a sustainable transportation fuel and then powering 4,500 natural gas vehicles and one commuter train. The increased use of biogas was seen as a key step toward creating a more sustainable society and a fossil-fuel-free future.[1]

The thermo-chemical decomposition of carbonaceous materials ("gasification"/ "pyrolysis") at elevated temperatures and in an oxygen-deprived environment

---

[1] http://www.businessregiongoteborg.com/newsarchives/newsarticles2014/gobigasbiogasplantina ugurated.5.5783eddf144e09ec4fef3f33.html, last accessed 1 April, 2014.

© Springer India 2015
S. Sunderasan, *Cleaner-Energy Investments*,
DOI 10.1007/978-81-322-2062-6_15

generated syngas (synthesis gas, a combination of hydrogen, carbon dioxide, and carbon monoxide). Since the fuels used were of recent biological origin, as with woodchips, tree bark, and other residues, from managed forests or organic waste, the gas produced in the reactor was considered renewable fuel and the power generated from its combustion was considered "sustainable."

Project Description

Goteborg Energi's Euro 150 million/20 MW (+80 MW) Gothenburg Biomass Gasification Project ("GoBiGas", Fig. 15.1) commissioned in late 2013 was designed to gasify 50,000 t of forest wastes and wood pellets per year and to produce gas for use as fuel in the transport sector. The project was a major bet on biogas and was designed to deliver gas to fuel 80,000–100,000 cars. The plant was believed to be the first of its kind anywhere in the world, and the gas generated was said to be similar to natural gas (Messenger 2014). The indirect gasification process employed at the plant was developed by carbon-neutral power plant specialists, *Repotec* of Austria (www.repotec.at). The equipment was supplied and installed by Finnish paper pulp and biowaste equipment major *Valmet* (formerly, *Metso*) (www.valmet. com). The plant was built, first, to demonstrate the possibilities with commercial scale gasification technology and, secondly, to build a plant to provide sustainable and carbon-neutral fuel (GoBiGas 2014). The project was to be run in collaboration with Germany-based E.ON, among the world's largest investor-owned electric utility service providers.

The heat from the hotbed material was transferred from the combustion chamber to the gasification reactor. The biomass feedstock was fed to the reactor where it came in contact with the heat and underwent thermo-chemical decomposition.

The indirect gasification process generated high calorific value gas, which, after cleaning and methanation, could be mixed with natural gas for use in gas-powered

Fig. 15.1 An aerial view of the GoBiGas project; reproduced with permission

vehicles. This combination of the biomass gasification process with a methanation plant was the first of its kind ever.

The plant was designed to generate 150 GWh energy equivalent of biogas, to run 15,000 cars (or 400 buses) a year. The project contributed toward achieving Sweden's 10 % renewable and climate-neutral transport fuels by the year 2020 including ethanol, biogas, and biofuels including fatty acid methyl esters (FAME) toward reducing carbon emissions and toward enhancing security of energy supplies. The plant was designed and developed by jointly by industry participants, the Chalmers University of Technology and the Swedish Government. Performance of the research-scale boiler was first tested at the University before contracting Valmet for commercial supply and installation.

Sweden had large tracts of sustainably managed forests within an active forest product industry and large volumes of agri-residue readily available for biofuel production. Fossil fuel substitution policies encouraged research and development into bioenergy. The Swedish Energy Agency had supported pilot plants for producing synthetic gas from biomass and ethanol from cellulose and from black liquor (Lewald 2014). Proposed "Phase 2" of the GoBiGas project was expected to range 80–100 MW, subject to lowering of capital expenditure, achieving higher overall efficiency, and developing better and cheaper gas cleaning processes (Hannula and Kurkela 2014).

Project Location

The Scandinavian region, in general, and Sweden, in particular, have always been at the forefront of research and technology development in biomass processing and gasification of wood residues to displace oil and petroleum derived fuels. In 1984, the 20 MW$_{th}$/146,300 Nm^3 per day *Norrsundet Bruks* plant was commissioned to convert biomass (mainly bark) and waste to syngas. In 1993, the 14MW$_{th}$ *Värnamo* IGCC demonstration plant was set up to generate electricity from biomass and waste (NETL 2014). The Gothenburg region had evolved into a clean-tech hub, employing some 9,000 persons. The region received the "Blue Sky Award" from the American non-profit organization, Calstart for the Biogas Väst initiative in 2008. The Calstart Blue Sky Award was designed to recognize leadership, commitment, and innovation for clean and sustainable transportation.[2] The plant was built in partial fulfillment of Goteborg Energi's vision to deliver 1 TWh of biogas before the year 2020, equivalent to the fuel consumed by 100,000 cars. The gas was to be piped through an existing natural gas grid in the southern and western parts of Sweden.

---

[2] http://www.calstart.org/events/calstart-events/Past-Event/Blue-Sky-Awards-Past.aspx, last accessed 1 April 2014.

The plant was located close to electricity, gas, and district heating network hubs in Gothenburg, while also allowing for convenient access for fuel and people by ship and rail. Water for plant operations was to be sourced from the Göta River.

Technology and Design

Goteborg Energi undertook a feasibility study involving Swedish and Dutch experts to compare candidate technologies: pressurized oxygen blown gasification and indirect gasification. Austrian firm *Repotec* and the indirect gasification technology were chosen in the year 2007 based on operational experience with the technology and its proven technical and economic performance. *Repotec* had built an 8-MW plant in Güssing, Austria, which had been in service since 2002, clocking over 60,000 h of operating history. At the same site, Swiss thermal waste treatment and gas cleaning specialists *Clean Technology Universe* (www.ctu.ch) had installed a pilot methanation plant. Operations personnel from the Güssing site were invited to evaluate the project design. The project was designed to convert 65 % of the biomass into biogas and to work for a minimum of 90 % (or about 7,900–8,000 h each year). The second stage of 80–100 MW_{gas} was scheduled for commissioning in the year 2016, subject to studying the operational performance of the first stage ($20MW_{gas}$).

Project Financing

The Euro 150 million initiative was part-funded by a grant of SEK 222 million ($\sim$24 million Euro) from the Swedish Energy Agency. Research and development continued at the Chalmers University of Technology to facilitate the construction of Phase 2 (80–100 MW/640–800 GWh) for which the European Union (EU) had pledged SEK 520 million ($\sim$58.8 million Euro). One TWh of biogas delivered by the consolidated project by the year 2020 would represent 25–30 % of the fuel consumed for transportation in all of Goteborg. The heat recovered from the reactor was to be used for district heating, and the ash was to be returned to forest lands to serve as natural fertilizer.

Positioning Biogas

- Fuel from the GoBiGas project had the lowest "well to wheels" impact on the environment at under 10 g CO_2eq/km, compared to gasoline vehicles at about 165, diesel vehicles at 145, and fossil methane at about 135 (Gunnarsson 2014).